International Conference on Exotic Atoms and Related Topics (EXA2021)

Held online

EPJ Web of Conferences Volume 262 (2022)

Vienna, Austria
13 - 17 September 2021

Editors:

Paul Buehler
Eberhard Widmann

ISBN: 978-1-7138-5228-5

Printed from e-media with permission by:

Curran Associates, Inc.
57 Morehouse Lane
Red Hook, NY 12571

Some format issues inherent in the e-media version may also appear in this print version.

For additional information, please contact EDP Sciences – Web of Conferences at the address below.

EDP Sciences – Web of Conferences
17, Avenue du Hoggar
Parc d'Activité de Courtabœuf
BP 112
F-91944 Les Ulis Cedex A
France

Phone: +33 (0) 1 69 18 75 75
Fax: +33 (0) 1 69 28 84 91

contact-edps@webofconferences.org

Additional copies of this publication are available from:

Curran Associates, Inc.
57 Morehouse Lane
Red Hook, NY 12571 USA
Phone: 845-758-0400
Fax: 845-758-2633
Email: curran@proceedings.com
Web: www.proceedings.com

TABLE OF CONTENTS

edp sciences

EPJ Web of Conferences

Statement of Peer review

In submitting conference proceedings to EPJ Web of Conferences, I certify to the Publisher that I adhere to the **Policy on Publishing Integrity** of the journal in order to safeguard good scientific practice in publishing.

1. All articles have been subjected to peer review administered by the proceedings editors.
2. Reviews have been conducted by expert referees, who have been requested to provide unbiased and constructive comments aimed, whenever possible, at improving the work.
3. Proceedings editors have taken all reasonable steps to ensure the quality of the materials they publish and their decision to accept or reject a paper for publication has been based only on the merits of the work and the relevance to the journal.

Title, date and place of the conference

International Conference on Exotic Atoms and Related Topics - EXA 2021

Proceedings editor(s):

Paul Buehler
Eberhard Widmann

Date and editor's signature

Vienna, April 7, 2022

17, avenue du Hoggar - PA de Courtabœuf – BP 112 - 91944 Les Ulis Cedex A (France)
Tél. : 33 (0)1 69 18 75 75 - Fax : 33(0)1 69 07 45 17 – www.edpsciences.org

Muonium Lamb shift: theory update and experimental prospects

Gianluca Janka[1], *Ben* Ohayon[1], and *Paolo* Crivelli[1,*]

[1]ETH Zurich, Institute for Particle Physics and Astrophysics, 8093 Zurich, Switzerland

Abstract. We review the theory of the Lamb shift for muonium, provide an updated numerical value and present the prospects of the Mu-MASS collaboration at PSI to improve upon their recent measurement. Due to its smaller nuclear mass, the contributions of the higher-order recoil corrections (160 kHz level) and nucleus self-energy (40 kHz level) are enhanced for muonium compared to hydrogen where those are below the level of the latest measurement performed by Hessels et al. and thus could not be tested yet. The ongoing upgrades to the Mu-MASS setup will open up the possibility to probe these contributions and improve the sensitivity of this measurement to searches for new physics in the muonic sector.

1 Introduction

Over the years, the $2S_{1/2} - 2P_{1/2}$ Lamb shift (LS) determination in hydrogen (H) has improved to the point that it became sensitive to the finite size effects of the nucleus [1]. To use the LS measurements to probe bound-state QED required an independent and accurate determination of the proton charge radius. Prior to 2010, the values for the proton charge radius determined from e-p scattering experiments and from hydrogen spectroscopy were in agreement [2]. With a new method based on muonic hydrogen laser spectroscopy, the proton charge radius was extracted with more than 10 times higher precision[3, 4]. Unexpectedly, the radius was found to be around 4 % smaller than from the other measurements, which resulted in a disagreement of about 7σ. This led to what was dubbed as the proton charge radius puzzle [5], triggering many experimental and theoretical efforts striving for a solution. The most recent determinations of the radius from e-p scattering [6] or from the Lamb shift [7] and the $2S - 4P$ transition [8] in hydrogen are agreeing with the value from muonic hydrogen, while the most recent measurement of the $2S_{1/2} - 8D_{5/2}$ transition in hydrogen is disagreeing by around 3σ [9]. Measurements of the $1S - 3S$ transition in hydrogen support both the smaller radius [10] as well as the originally accepted value [11]. Although the puzzle is arguably coming to an end, the discrepancy between the experiments is yet to be understood [12].

Muonium (M), which is composed of a positively charged muon orbited by an electron, is an excellent candidate to probe bound-state QED due to its lack of internal structure and therefore freeing it from finite size effects. Furthermore, M is lighter than hydrogen, making it more suitable for testing effects related to its nucleus mass such as the recoil or the nucleus

*e-mail: crivelli@phys.ethz.ch

self-energy. The last numerical value of the M LS calculated with bound-state QED was by J. R. Sapirstein and D. R. Yennie [13] to be at 1047.49(1)(9) MHz in 1990. Here we provide an updated value which takes into account the improvements in the LS calculations done in the last three decades.

2 Lamb shift calculation

To calculate the energy difference between the $2S_{1/2}$ and $2P_{1/2}$ states, we followed the theory of the hydrogen atom as summarized by CODATA 2018 [14] and added some relevant terms for muonium which due to their smallness in hydrogen were neglected. In the following, we use m_e for the electron, m_n for the nuclear and m_r for the reduced mass. The reduced mass is defined as:

$$m_r = \frac{m_e m_n}{m_e + m_n} \tag{1}$$

We first tabulate the non-recoil contributions, which are identical between muonium and hydrogen up to powers of $\left(\frac{m_r}{m_e}\right)$. The main contribution to the Lamb shift is the self-energy E_{SE}, where the bound electron emits and reabsorbs a virtual photon. A similar effect comes from the vacuum polarization, which can be looked at as an effect from the self-energy of the gauge boson, where it produces a virtual e^+e^- (E_{VP}), or in more suppressed cases $\mu^+\mu^-$ ($E_{\mathrm{VP}\mu}$) or hadronic ($E_{\mathrm{VP,had}}$) pair, which eventually annihilate again. The contributions can be calculated as:

$$E_{\mathrm{SE}} = \frac{\alpha}{\pi}\frac{(Z\alpha)^4}{n^3}\left(\frac{m_r}{m_e}\right)^3 F(Z\alpha)\, m_e c^2 \tag{2}$$

$$E_{\mathrm{VP}} = \frac{\alpha}{\pi}\frac{(Z\alpha)^4}{n^3}\left(\frac{m_r}{m_e}\right)^3 H(Z\alpha)\, m_e c^2 \tag{3}$$

$$E_{\mathrm{VP}\mu} = \frac{\alpha}{\pi}\frac{(Z\alpha)^4}{n^3}\left(\frac{m_r}{m_e}\right)^3 \left(\frac{m_e}{m_\mu}\right)^2 \left[-\frac{4}{15}\right]\delta_{l0}\, m_e c^2 \tag{4}$$

$$E_{\mathrm{VP,had}} = 0.671(15)\, E_{\mathrm{VP}\mu} \tag{5}$$

$$F(Z\alpha) = L\,A_{41} + A_{40} + (Z\alpha)A_{50} + (Z\alpha)^2\left[L^2 A_{62} + L\,A_{61} + G_{\mathrm{SE}}(Z\alpha)\right] \tag{6}$$

$$H(Z\alpha) = V_{40} + (Z\alpha)V_{50} + (Z\alpha)^2\left[L\,V_{61} + G_{\mathrm{VP}}(Z\alpha) + G^{(R)}_{\mathrm{VP}}(Z\alpha)\right], \tag{7}$$

where we denote $L = \ln\left(\frac{m_e}{m_r}\frac{1}{(Z\alpha)^2}\right)$ and δ_{ij} as the Kronecker delta.

Higher-order nonrecoil contributions result from two- and three-photon corrections $E_{2\mathrm{ph}}$ and $E_{3\mathrm{ph}}$:

$$E_{2\mathrm{ph}} = \left(\frac{\alpha}{\pi}\right)^2 \frac{(Z\alpha)^4}{n^3}\left(\frac{m_r}{m_e}\right)^3 F_{2\mathrm{ph}}(Z\alpha)\, m_e c^2 \tag{8}$$

$$E_{3\mathrm{ph}} = \left(\frac{\alpha}{\pi}\right)^3 \frac{(Z\alpha)^4}{n^3}\left(\frac{m_r}{m_e}\right)^3 F_{3\mathrm{ph}}(Z\alpha)\, m_e c^2 \tag{9}$$

$$\begin{aligned} F_{2\mathrm{ph}}(Z\alpha) &= B_{40} + (Z\alpha)B_{50} + (Z\alpha)^2\left[L^3 B_{63} + L^2 B_{62} + L\,B_{61} + B_{60}\right] \\ &\quad + (Z\alpha)^3\left[L^2 B_{72} + L\,B_{71} + \ldots\right] \end{aligned} \tag{10}$$

$$F_{3\mathrm{ph}}(Z\alpha) = C_{40} + (Z\alpha)C_{50} + (Z\alpha)^2\left[L^3 C_{63} + L^2 C_{62} + L\,C_{61} + C_{60} + \ldots\right]. \tag{11}$$

Compared to hydrogen, the nucleus of muonium is approximately nine times lighter, making it more sensitive to recoil corrections. We list the recoil corrections appearing in CODATA 2018 such as the Barker-Glover correction $E_{\rm BKG}$, the relativistic recoil $E_{\rm rec,S} + E_{\rm rec,R}$, the radiative recoil $E_{\rm RR}$ and the nucleus self-energy $E_{\rm SEN}$ and later additional contributions potentially important to M:

$$E_{\rm BKG} = \frac{(Z\alpha)^4}{2n^3}\left(\frac{m_r}{m_e}\right)^3\left(\frac{m_e}{m_n}\right)^2\frac{1-\delta_{l0}}{\kappa\,(2l+1)}m_ec^2 \tag{12}$$

$$E_{\rm rec,S} = \frac{(Z\alpha)^5}{\pi n^3}\left(\frac{m_r}{m_e}\right)^3\left(\frac{m_e}{m_n}\right)S(Z\alpha)\,m_ec^2 \tag{13}$$

$$E_{\rm rec,R} = \frac{(Z\alpha)^6}{n^3}\left(\frac{m_r}{m_e}\right)^3\left(\frac{m_e}{m_n}\right)R(Z\alpha)\,m_ec^2 \tag{14}$$

$$E_{\rm RR} = \frac{\alpha}{\pi}\frac{(Z\alpha)^5}{\pi n^3}\left(\frac{m_r}{m_e}\right)^3\left(\frac{m_e}{m_n}\right)Q(Z\alpha)\,\delta_{l0}\,m_ec^2 \tag{15}$$

$$E_{\rm SEN} = \frac{Z^2\alpha}{\pi}\frac{(Z\alpha)^4}{n^3}\left(\frac{m_r}{m_e}\right)^3\left(\frac{m_e}{m_n}\right)^2\left[\left(\frac{\mathbf{10}}{\mathbf{9}}+\frac{4}{3}\ln\left[\frac{m_n}{m_r(Z\alpha)^2}\right]\right)\delta_{l0}-\frac{4}{3}\ln k_0\right]m_ec^2 \tag{16}$$

$$S(Z\alpha) = L\,D_{51} + D_{50} - \frac{2\delta_{l0}}{m_n^2-m_e^2}\left[m_n^2\ln\left(\frac{m_e}{m_r}\right)-m_e^2\ln\left(\frac{m_n}{m_r}\right)\right] \tag{17}$$

$$R(Z\alpha) = D_{60} + (Z\alpha)\,G_{\rm REC}(Z\alpha) \tag{18}$$

$$Q(Z\alpha) = 6\zeta[3] - 2\pi^2\ln(2) + \frac{35\pi^2}{36} - \frac{448}{27} + (Z\alpha)\left[\frac{2}{3}\pi\ln\left((Z\alpha)^{-2}\right)^2\right] + ... \tag{19}$$

where κ is the angular momentum parity quantum number, ζ is the Riemann zeta function and $\ln k_0$ is the relevant Bethe logarithm. In atoms with a nuclear size such as hydrogen, the $\frac{10}{9}$ term in $E_{\rm SEN}$ (Eq. 16) would be absorbed in the finite size contributions. Since muonium has no sub-structure, it is free from finite size effects and therefore the finite nuclear size contributions can be omitted but $E_{\rm SEN}$ has to be extended. See also the discussion in Ref. [15].

All the coefficients A, B, C, D and V, the remainder functions G as well as the uncertainties of each contribution are tabulated in Ref. [14, 16–18]. These quantities typically depend on the quantum numbers n (principal), l (azimuthal) and j (total angular momentum). Their notation is suppressed here for brevity.

In addition to the above, we collect higher-orders recoil contributions. These are negligible for H, but are up to two orders of magnitude larger for M, and, thus, should be considered. We first focus on radiative-recoil contributions. Czarnecki and Melnikov [19] have calculated higher-order mass contributions to the radiative insertions in the electron line, of which the 2$^{\rm nd}$ order is

$$E_{\rm RR2e} = \frac{\alpha}{\pi}\frac{(Z\alpha)^5}{n^3}\left(\frac{m_r}{m_e}\right)^3\left(\frac{m_e}{m_n}\right)^2\left[\pi 8\ln(2) - \pi\frac{127}{32}\right]\delta_{l0}\,m_ec^2, \tag{20}$$

which returns a negligible 80 Hz for the M LS. Another radiative recoil contribution comes from polarization insertions calculated in [20], whose 2$^{\rm nd}$ order in mass correction is given in [21]

$$E_{\rm RR2p} = \frac{\alpha}{\pi}\frac{(Z\alpha)^5}{n^3}\left(\frac{m_r}{m_e}\right)^3\left(\frac{m_e}{m_n}\right)^2\left[-\frac{3\pi}{16}\right]\delta_{l0}\,m_ec^2, \tag{21}$$

returning a negligible −10 Hz for the M LS. Very recently, spin-independent three-loop radiative corrections have been calculated for positronium and muonium [21], giving

$$E_{\mathrm{RR3}} = \left(\frac{\alpha}{\pi}\right)^2 \frac{(Z\alpha)^5}{\pi n^3} \left(\frac{m_r}{m_e}\right)^3 \left(\frac{m_e}{m_n}\right) [-11.4...]\ \delta_{l0}\ m_e c^2, \tag{22}$$

which returns a negligible −30 Hz contribution to the M LS.

For pure recoil contributions, equations 13 and 17 account for recoil terms up to order $(Z\alpha)^5(m_e/m_n)^3 \ln(m_e/m_n)$. The expansion in mass ratio of the pure recoil term of order $(Z\alpha)^6$ is calculated partially in [19], which to 2nd order gives

$$E_{\mathrm{rec,R2}} = \frac{(Z\alpha)^6}{n^3} \left(\frac{m_r}{m_e}\right)^3 \left(\frac{m_e}{m_n}\right)^2 \left[\frac{4}{\pi^2} \ln\frac{m_n}{m_e} - \frac{8}{3} \ln\frac{m_n}{m_e} - \frac{12\zeta_3}{\pi^2} + \frac{3}{\pi^2} + \frac{8}{3}\right] \delta_{l0}\ m_e c^2, \tag{23}$$

returning −575 Hz for the M LS. This contribution is not negligible compared with the 0.8 kHz uncertainty attached to equation 15 and 19. To account for it being partial, we take the full contribution as its uncertainty. The total uncertainty in the muonium lamb shift is thus 1.0 kHz.

As another addition to CODATA 2018, we include the off-diagonal hyperfine-structure contribution E_{HFS}. This additional recoil contribution arises from the mixing of the fine structure sublevels due to the hyperfine interaction as given by for example V. Yerokhin et al. [16]:

$$E_{\mathrm{HFS}} = \frac{\alpha^2(Z\alpha)^2}{n^3} \left(\frac{m_e}{m_p}\right)^2 \left(\frac{\mu}{\mu_{\mathrm{N}}}\right)^2 \frac{2I(I+1)}{81} (-1)^{j+1/2} \delta_{l1} m_e c^2. \tag{24}$$

The I is the nuclear spin quantum number, μ the nuclear magnetic moment and μ_{N} the nuclear magneton.

For both hydrogen and muonium, we evaluate these contributions for the $2S_{1/2}$ and $2P_{1/2}$ levels. Their difference in frequency are summarized in Tab. 1. We calculate an updated LS frequency for muonium while validating our result with the values summarized by Yerokhin for hydrogen [16]. The finite-size and other nuclear effects for hydrogen are not shown, but would amount to around 138 kHz. Only uncertainties ≥ 500 Hz are given. Our final value for the LS in M is 1047.498(1) MHz. It is in agreement and 100 times more accurate than the latest bound-state QED calculation by Sapirstein and Yennie [13]. However, C. Frugiuele et al. recently calculated the M LS using effective field theory to be 1047.284(2) MHz [22], which differs around 200 kHz from our value and, thus, deserves further investigations.

Comparing the contributions for hydrogen and muonium, we identify the Barker-Glover correction as well as the nucleus self-energy as the two most interesting contributions for future experiments. The current best determination of the hydrogen Lamb shift has an uncertainty of around 3 kHz [7], hence these two contributions could not be resolved so far. In muonium, they are enhanced and can be probed by achieving Lamb shift uncertainties below 160 kHz and 40 kHz, respectively.

3 Mu-MASS experiment

The Mu-MASS experiment was proposed in 2018 [23] with the main goal to improve upon the $1S-2S$ transition in muonium. Key elements of the experiment are the unique low energy muon (LEM) beam line at PSI, the efficient tagging of the incoming muons, the production of M at cryogenic temperatures [24], a high-power CW UV-laser for excitation [25] as well as an efficient detection of the metastable M(2S) atoms. To test the detection scheme, an intense

Table 1. Summary of the calculated contributions to the hydrogen and muonium Lamb shift transition. The theoretical calculations from Sapirstein & Yennie [13] and from Frugiuele et al. [22] are included for comparison. Uncertainties smaller than 0.5 kHz are not tabulated.

	Largest Order	Hydrogen (MHz)	Muonium (MHz)
E_{SE}	$\alpha\,(Z\alpha)^4\,L$	1084.128	1070.940
E_{VP}	$\alpha\,(Z\alpha)^4$	−26.853	−26.510
$E_{VP\mu+had}$	$\alpha\,(Z\alpha)^4(m_e/m_\mu)^2$	−0.001	−0.001
E_{2ph}	$\alpha^2(Z\alpha)^4$	0.065	0.065
E_{3ph}	$\alpha^3(Z\alpha)^4$	0.000	0.000
E_{BKG}	$(Z\alpha)^4\ (m_e/m_n)^2$	−0.002	−0.168
$E_{rec,S}$	$(Z\alpha)^5\,L\,(m_e/m_n)$	0.358	3.138
$E_{rec,R}$	$(Z\alpha)^6\ (m_e/m_n)$	−0.001	−0.012
$E_{rec,R2}$	$(Z\alpha)^6\ (m_e/m_n)^2$	−0.000	−0.001(1)
E_{RR}	$\alpha\,(Z\alpha)^5\ (m_e/m_n)$	−0.002	−0.014(1)
E_{RR2e+p}	$\alpha\,(Z\alpha)^5\ (m_e/m_n)^2$	0.000	0.000
E_{RR3}	$\alpha^2(Z\alpha)^5\ (m_e/m_n)$	−0.000	−0.000
E_{SEN}	$Z^2\alpha\,(Z\alpha)^4\,(m_e/m_\mu)^2$	0.001	0.041
E_{HFS}	$\alpha^2(Z\alpha)^2\,(m_e/m_n)^2$	0.002	0.019
Sum			1047.498(1)
Ref. [13]			1047.49(9)
Ref. [22]			1047.284(2)

M(2S) beam of the order of 100 Hz was formed, opening up the possibility to improve upon the muonium Lamb shift [26].

The metastable beam is produced at the LEM beamline by guiding a continuous 10 keV μ^+ beam on a thin carbon foil ($\approx$ 10 nm). While passing through the foil, the μ^+ can pick up an electron to form M, predominately in the ground state, but also 5 % to 10 % in the excited $2S$ state. During this process, the μ^+ also releases secondary electrons, which are detected and used for tagging.

The M(2S) passes through two microwave regions, one to depopulate unwanted hyperfine states and the other to drive the Lamb shift transition $2S$-$2P_{1/2}$ of interest. Once reaching the $2P$ state, the atom relaxes back to the ground state with a lifetime of 1.6 ns. The beam reaches then the detection chamber, where an electric field of the order of 300 V cm^{-1} is quenching the remaining $2S$ atoms. The emitted Lyα photons (wavelength of 122 nm) are detected by two coated MCP detectors. The muonium beam is eventually stopping on an MCP at the end of the beamline, creating the stop signal. The experimental sketch is shown in Fig. 1. By demanding a signal in all three detectors in specific time windows, the amount of Lyα photons detected at the applied microwave frequency can be extracted. At the frequency where the minimal amount of Lyα was detected, the microwave was optimally tuned and there-

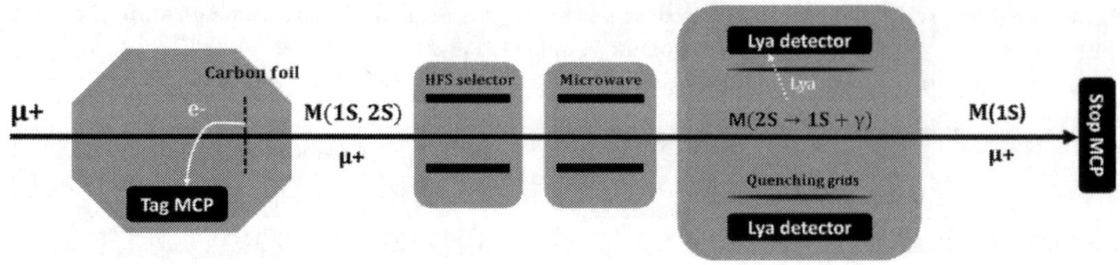

Figure 1. The experimental sketch of the muonium Lamb shift measurement within the Mu-MASS experiment at PSI.

fore depopulating the metastable beam most efficiently already before reaching the detection chamber.

The M Lamb shift was determined to be 1047.2(25) MHz [27], where the uncertainty is almost completely of statistical origin. This result is an improvement of an order of magnitude compared to the last best determinations [28, 29]. As a next step, the foil thickness will be reduced from 10 nm to a few layers of graphene [30, 31] (~1 nm), which will increase our detectable M(2S) rate by a factor ranging from 15 to 25.

4 Conclusions

Calculating each contribution separately shows that when reaching an uncertainty of around 160 kHz in muonium, one becomes sensitive to higher-order recoil effects such as E_{BKG}, which cannot be resolved yet with hydrogen. In the context of Mu-MASS, by replacing the 10 nm-thick carbon foil for producing muonium with a few layers of graphene, an uncertainty of 160 kHz would be feasible within around a week of beamtime. Lowering the uncertainty further to 40 kHz, the effect coming from the nucleus self energy E_{SEN} can be studied, which is also not in reach right now with hydrogen. The ongoing development of the muCool project [32] at PSI will improve the beam quality and provide an order of magnitude larger muon flux, allowing us to achieve such uncertainty. The realization of the High Intensity Muon Beam (HIMB) project at PSI [33] would allow to push this even further as summarized in Tab. 2.

Table 2. Summary of possible upgrades to the Mu-MASS setup and which contributions of the M LS could be probed.

Beamline	Target	Timeline	M(2S) (Hz)	LS Uncertainty (kHz / 10 d)	Contributions
PiE4/LEM	C-Foil	2021	5	1000	E_{SE}, E_{VP}, $E_{rec,S}$
PiE4/LEM	Graphene	2023	100	200	E_{BKG}
PiE1/muCool	Graphene	2025	1000	70	E_{2ph}
PiE1/muCool	Gas	2026	5000	30	E_{SEN}
HiMB/muCool	Gas	2029	100 000	10	E_{RR}, E_{HFS}, $E_{rec,R}$

Acknowledgements

This work is supported by the ERC consolidator grant 818053-Mu-MASS and the Swiss National Science Foundation under the grant 197346. BO acknowledges support from the European Union's Horizon 2020 research and innovation program under the Marie Skłodowska-Curie grant agreement No. 101019414. We are very grateful to K. Pachucki for his help with the theoretical calculations, S. Karshenboim, M. Eides and A. Czarnecki for the very useful discussions and T. Udem for sharing his code for the hydrogen calculations.

References

[1] S.R. Lundeen, F.M. Pipkin, Phys. Rev. Lett. **46**, 232 (1981)

[2] P.J. Mohr, D.B. Newell, B.N. Taylor, Rev. Mod. Phys. **88**, 035009 (2016)

[3] R. Pohl et al., Nature **466**, 213 (2010)

[4] A. Antognini, F. Nez, K. Schuhmann, F.D. Amaro, F. Biraben, J.M.R. Cardoso, D.S. Covita, A. Dax, S. Dhawan, M. Diepold et al., Science **339**, 417 (2013)

[5] R. Pohl, R. Gilman, G.A. Miller, K. Pachucki, Annual Review of Nuclear and Particle Science **63**, 175 (2013)

[6] W. Xiong, A. Gasparian, H. Gao, D. Dutta, M. Khandaker, N. Liyanage, E. Pasyuk, C. Peng, X. Bai, L. Ye et al., Nature (London) **575** (2019)

[7] N. Bezginov, T. Valdez, M. Horbatsch, A. Marsman, A.C. Vutha, E.A. Hessels, Science **365**, 1007 (2019)

[8] A. Beyer, L. Maisenbacher, A. Matveev, R. Pohl, K. Khabarova, A. Grinin, T. Lamour, D.C. Yost, T.W. Hänsch, N. Kolachevsky et al., Science **358**, 79 (2017)

[9] A.D. Brandt, S.F. Cooper, C. Rasor, Z. Burkley, D.C. Yost, A. Matveev, *Measurement of the* $2s_{1/2} - 8d_{5/2}$ *transition in hydrogen* (2021), 2111.08554

[10] A. Grinin, A. Matveev, D.C. Yost, L. Maisenbacher, V. Wirthl, R. Pohl, T.W. Hänsch, T. Udem, Science **370**, 1061 (2020)

[11] H. Fleurbaey, S. Galtier, S. Thomas, M. Bonnaud, L. Julien, F.m.c. Biraben, F.m.c. Nez, M. Abgrall, J. Guéna, Phys. Rev. Lett. **120**, 183001 (2018)

[12] J.P. Karr, D. Marchand, E. Voutier, Nature Reviews Physics pp. 1–14 (2020)

[13] J.R. Sapirstein, D.R. Yennie, Adv. Ser. Direct. High Energy Phys. **7**, 560 (1990)

[14] E. Tiesinga, P.J. Mohr, D.B. Newell, B.N. Taylor, Journal of Physical and Chemical Reference Data **50**, 033105 (2021)

[15] K. Pachucki, Phys. Rev. A **52**, 1079 (1995)

[16] V.A. Yerokhin, K. Pachucki, V. Patkóš, Annalen der Physik **531**, 1800324 (2019)

[17] S.G. Karshenboim, A. Ozawa, V.G. Ivanov, Phys. Rev. A **100**, 032515 (2019)

[18] S.G. Karshenboim, A. Ozawa, V.A. Shelyuto, R. Szafron, V.G. Ivanov, Physics Letters B **795**, 432 (2019)

[19] I. Blokland, A. Czarnecki, K. Melnikov, Phys. Rev. D **65**, 073015 (2002)

[20] M.I. Eides, H. Grotch, Phys. Rev. A **52**, 1757 (1995)

[21] M.I. Eides, V.A. Shelyuto, *Three-loop corrections to lamb shift in muonium and positronium* (2021), 2110.13279

[22] C. Frugiuele, J. Pérez-Ríos, C. Peset, Phys. Rev. D **100**, 015010 (2019)

[23] P. Crivelli, Hyperfine Interactions **239** (2018)

[24] A. Antognini, P. Crivelli, T. Prokscha, K.S. Khaw, B. Barbiellini, L. Liszkay, K. Kirch, K. Kwuida, E. Morenzoni, F.M. Piegsa et al., Phys. Rev. Lett. **108**, 143401 (2012)

[25] Z. Burkley, L. de Sousa Borges, B. Ohayon, A. Golovozin, J. Zhang, P. Crivelli, Opt. Express **29**, 27450 (2021)

[26] G. Janka, B. Ohayon, Z. Burkley, L. Gerchow, N. Kuroda, X. Ni, R. Nishi, Z. Salman, A. Suter, M. Tuzi et al., Eur. Phys. J. C **80**, 804 (2020)

[27] B. Ohayon, G. Janka, I. Cortinovis, Z. Burkley, L. de Sousa Bourges, E. Depero, A. Golovizin, X. Ni, Z. Salman, A. Suter et al., *Precision measurement of the lamb shift in muonium* (2021), 2108.12891

[28] K.A. Woodle, A. Badertscher, V.W. Hughes, D.C. Lu, M.W. Ritter, M. Gladisch, H. Orth, G. zu Putlitz, M. Eckhause, J. Kane et al., Phys. Rev. A **41**, 93 (1990)

[29] C.J. Oram, J.M. Bailey, P.W. Schmor, C.A. Fry, R.F. Kiefl, J.B. Warren, G.M. Marshall, A. Olin, Phys. Rev. Lett. **52**, 910 (1984)

[30] F. Allegrini, R.W. Ebert, S.A. Fuselier, G. Nicolaou, P.V. Bedworth, S.W. Sinton, K.J. Trattner, Optical Engineering **53**, 1 (2014)

[31] R.W. Ebert, F. Allegrini, S.A. Fuselier, G. Nicolaou, P. Bedworth, S. Sinton, K.J. Trattner, Review of Scientific Instruments **85**, 033302 (2014)

[32] A. Antognini, D. Taqqu, SciPost Phys. Proc. **5**, 030 (2021)

[33] M. Aiba et al. (2021), 2111.05788

EPJ Web of Conferences **262**, 01002 (2022)
EXA2021
https://doi.org/10.1051/epjconf/202226201002

A Rydberg hydrogen beam for studies of stimulated deexcitation

Tim Wolz[1,*], *Maxime* Allemand[2], *Daniel* Comparat[3], *Jules* Cras[3], *Carina* Killian[4], *Chloé* Malbrunot[1,**], *Fredrik* Parnefjord Gustafsson[4], *Martin* Simon[4], *Christophe* Siour[3], and *Eberhard* Widmann[4] on behalf of the AEGIS and ASACUSA-CUSP collaboration

[1]Physics Department, CERN
[2]École normale supérieure Paris-Saclay
[3]Université Paris-Saclay, CNRS, Laboratoire Aimé Cotton, Paris
[4]Stefan Meyer Institute, Vienna

Abstract. We present a Rydberg hydrogen beamline developed to commission techniques of stimulated deexcitation for application in antihydrogen experiments at CERN's Antiproton Decelerator. The stimulation of spontaneous decay is a key technology to enhance the number of ground-state anti-atoms available in a beam toward precision spectroscopy and gravity measurements.

1 Introduction

Atomic antimatter systems are synthesized at CERN's Antiproton Decelerator facility to perform stringent tests of CPT symmetry and gravity. Antihydrogen ($\overline{\mathrm{H}}$) atoms can be currently produced relying on a resonant charge-exchange of laser excited positronium (a short-lived bound state of an electron and a positron (e^+)) and trapped antiprotons ($\overline{p}$) [1]. Alternatively, experiments rely on a three-body-reaction in a $\overline{p}$-e^+ plasma involving an antiproton and two positrons, one of which falls into a bound state with the antiproton and the other one carries away the atomic binding energy [2, 3]. In both cases, $\overline{\mathrm{H}}$ atoms are formed in a distribution of highly excited Rydberg quantum states exhibiting radiative lifetimes up to several milliseconds. In particular the numerous high angular momentum states are very long-lived while experiments require $\overline{\mathrm{H}}$ atoms in their ground state.

Unless employing neutral atom traps, experiments cannot rely on slow spontaneous emission to obtain ground state antihydrogen [4]. It is thus of paramount importance to either initially form strongly bound $\overline{\mathrm{H}}$ quantum states (by for example acting, in the case of a three-body-reaction, on the $\overline{p}$-e^+ plasma parameters as discussed in [5]) or enhance the decay of nascent Rydberg $\overline{\mathrm{H}}$ states in current experimental conditions. In view of the latter approach, several deexcitation schemes relying on either electric and magnetic field [6, 7] or light mixing of Rydberg states [8] associated with, in most cases, laser stimulated deexcitation have been theoretically identified. The techniques allow in principle to achieve close to unity ground state fractions of initially populated levels with principal quantum numbers $n \sim 30$ within a few tens of microseconds.

We discuss here the concept and status of a hydrogen proof-of-principle experiment to commission stimulated deexcitation techniques for application in antihydrogen experiments.

*e-mail: tim.wolz@cern.ch
**Present address: TRIUMF, 4004 Wesbrook Mall, Vancouver, BC, V6T 2A3, Canada

2 Hydrogen proof-of-principle beamline

Due to the scarcity of antihydrogen atoms we have developed and built a hydrogen beamline to test and commission deexcitation techniques for application in experiments located at the Antiproton Decelerator. The experimental setup and different avenues toward the production of an excited Rydberg beam are discussed in the following.

2.1 Atomic hydrogen beamline

The setup consists of a discharge plasma source to produce a thermal beam of atomic hydrogen [9]. For this purpose, ultra-pure molecular hydrogen gas is produced from deionized water with an electrolysis generator. The gas is guided through a supply line consisting of a small buffer reservoir and a flow-controller from where it reaches, further downstream, a cylindrical quartz tube that is encased by a metallic resonator cavity. The latter (design #5 in [10]) can be hand-tuned to efficiently sustain a microwave discharge plasma at some tens of watts allowing to dissociate the flowing molecular gas. Hydrogen atoms are emitted through a small pinhole into the downstream vacuum region. A cryogenic beam shield that is cooled with compressed helium to temperatures around 25 K prevents thermal radiation from impacting the quantum state distribution in the atomic beam which can be probed further

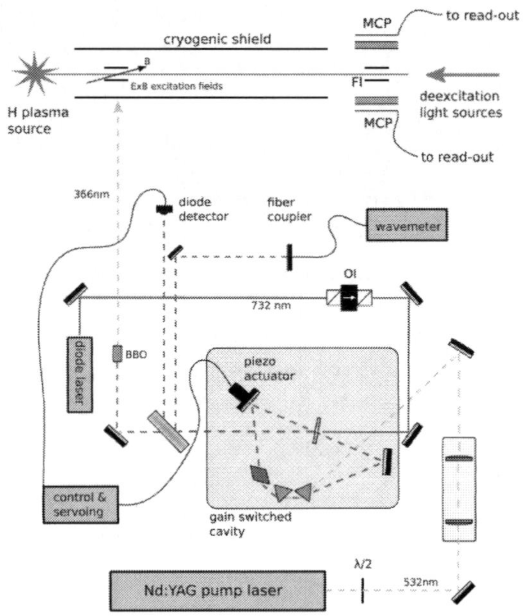

Figure 1: Top: Schematic of the hydrogen beamline including the electric field ionizer (FI) and the microchannel plate (MCP) particle detectors. The electric and magnetic fields required for optical Rydberg excitation (cf. section 2.2.1) yet need to be experimentally implemented. Bottom: Schematic of the injection seeded Ti:Sa Rydberg excitation laser.

down the atoms' flight path. Upon exiting from the thermal shield, the beam enters an electric field region generated by two ionization meshes that are mounted parallel to the beam propagation direction at a distance of 5 mm. The ionization products are accelerated toward and collected into the nearby MCP chevron stacks. The quantum state distribution of the beam is investigated by counting the ionization events per time interval as a function of a varied electric field ionization strength. Rydberg state mixing and deexcitation light sources can illuminate the atomic sample through a dedicated vacuum window at the rear-end of the setup. An illustration of the beamline is provided in the top part of Fig. 1. A photograph of the installation is shown in Fig. 2.

2.2 Hydrogen Rydberg beam production

The source produces in large majority atoms in ground-state. In order to develop deexcitation techniques it is thus required to excite a substantial fraction of the atomic beam toward a Rydberg state. We follow different avenues first of which is a collisional-optical excitation as discussed in subsection 2.2.1. Alternatively, Rydberg levels can be populated via recombination processes and electron impact inside the discharge plasma. We present, in section 2.2.2, results of a scan of the quantum state distribution of the beam emitted from the hydrogen source and discuss the complementarity of both approaches for our purpose.

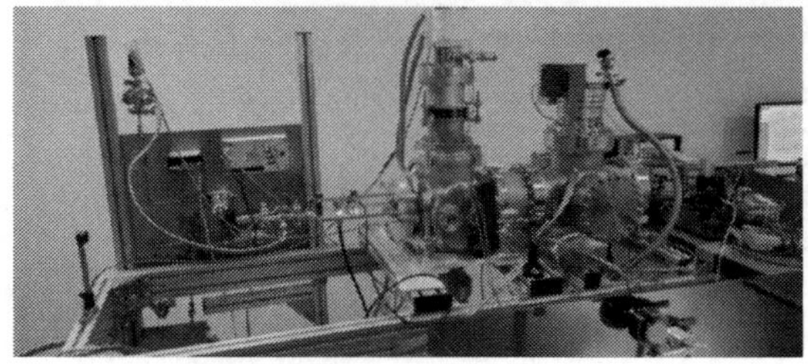

Figure 2: Photograph of the hydrogen beamline. The molecular gas supply and the pink plasma discharge sustained by the Evenson cavity are shown on the left. Further downstream the beam, two vacuum chambers host the cryogenic beam shield and the instrumentation required to probe the beam's quantum state distribution, respectively.

2.2.1 Collisional-optical excitation

We have developed and commissioned a laser capable of exciting 2s metastable atoms (radiative lifetime of 0.12 s) to highly excited Rydberg states with $n \sim 30$. The setup is inspired by the work presented in [11]. A commercial frequency doubled Nd:YAG laser provides light pulses with a width of 10 ns and maximum average pulse energies of 200 mJ at 532 nm. The pump beam is guided onto a Ti:Sa crystal inside a gain switched cavity. A fraction of the resulting red laser light impinges on a detection diode and a wavemeter while the main beam is again frequency doubled within a BBO crystal to obtain the required $2s \rightarrow 30p$ transition wavelength of 366 nm. In order to achieve a narrow spectral emission profile, the cavity is injection seeded with a few mW cw laser diode at the desired wavelength. The cavity is kept at resonance with the seeding beam relying on a piezo-electric mirror mount to compensate for drifts and therefore establish stable single-mode emission. The piezo-element is controlled with an electronic box and associated software that is based on the work presented in [12]. The installation is illustrated in the bottom part of Fig. 1 and a picture is shown in Fig. 3. The parameters of the optical setup and its light output characteristics at the mentioned wavelengths are specified in Table 1. In the unseeded case, the light emission occurs in both directions along the resonator cavity, whereas the pulse energy in single-mode operation is concentrated into the forward direction only. This results in an approximate doubling of the output average pulse energy when the cavity is injection seeded. The maximum achievable 366 nm average pulse energy is found to be 10.1(9) mJ. This parameter is limited by the damage threshold of the Ti:Sa crystal which is reached at ~ 30 mJ average pulse energy in the 732 nm beam just downstream the cavity.

The techniques proposed for stimulated deexcitation allow to address the entire distribution of quantum states formed in a typical antihydrogen experiment – most importantly including the preferentially populated and longest-lived high (n, l, m) levels with lifetimes $\tau \propto n^3 l^2$. We are thus aiming at producing a beam of Rydberg atoms containing these states of interest. To this goal, we are in the process of designing coils and electrodes to allow for

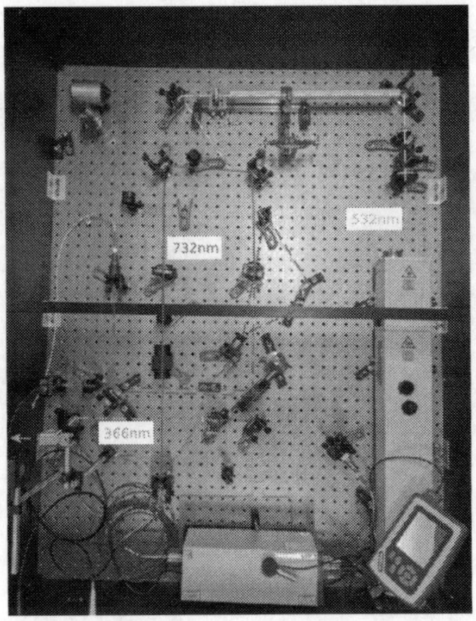

Figure 3: Photograph of the Rydberg excitation laser setup. The laser beam paths are shown in analogy to the illustration shown in Fig. 1. The picture is reprinted from [13].

pulse repetition rate	10 Hz
pulse duration	10 ns
cavity length	0.65 m
output coupler reflectivity	20 %
Ti:Sa crystal dimensions	5 x 10 x 20 mm^3
BBO crystal dimensions	5 x 5 x 10 mm^3
beam waist (532/732/366 nm)	3.5/2/2 mm
injection diode cw power	5 mW
injection diode beam diameter	2 mm
single-mode locking efficiency	97 %

	Average laser pulse energy [mJ]		
532 nm	107.2 ± 1.3	127.3 ± 1.3	145.1 ± 1.3
732 nm	**14.7 ± 1.4**	**22.5 ± 2.4**	**29.6 ± 3.9**
366 nm	1.7 ± 0.2	2.9 ± 0.2	4.1 ± 0.3
	3.9 ± 0.6	**7.1 ± 0.8**	**10.1 ± 0.9**

Table 1: Top: Summary of the parameters of the optical installation. Bottom: Average laser pulse energies for the pump beam at 532 nm, the Ti:Sa cavity output at 732 nm and the frequency doubled BBO laser output at 366 nm. The values are displayed in bold font for the injection seeded and in regular font for the unseeded case, respectively.

optical excitation, using the laser discussed above, toward circular states via the crossed fields method discussed and experimentally demonstrated in [14, 15].

The optical transition stimulated by the laser discussed above requires a metastable 2s beam which can be generated from ground-state relying on different approaches. Optical excitation toward the 3p levels allows to obtain 2s states according to the spontaneous decay branching ratio of 12 % [16]. Another possibility lies in a two-photon 1s → 2s excitation or electron collisional processes [17, 18]. For the latter technique we have developed and currently commission an electron gun. The beamline is designed such that the resulting hydrogen beam deflections caused by electron impact can be compensated for.

Following the observation of Rydberg states produced through collisional and recombination processes inside the electron discharge plasma, as discussed in the following part, metastable 2s atoms can be expected to emerge either from the source through electron collisions inside the plasma or population through spontaneous decay from higher excited states downstream the plasma inside the beam. We could, as of now, see no clear evidence of the 2s state being populated in the beam when detecting Lyman-alpha photons with the MCPs and relying on electric field quenching. Possible explanations involve the depopulation via 2p states already close to the plasma within the microwave cavity (indeed the few tens of W required to sustain the discharge on a few cm result in electric fields of some ten V cm^{-1}) or just upstream the field-ionization region caused by stray fields of the MCPs which would result in a reduced detection efficiency.

2.2.2 Emission of excited states from the electron discharge plasma

We performed simulations relying on the collisional radiative model Yacora [19] to theoretically investigate the production, in the microwave discharge, of highly excited Rydberg states that are of interest for studies of stimulated deexcitation. We estimate typical electron densities in our setup to be $10^{14}\,\mathrm{m}^{-3} \leq n_e \leq 10^{15}\,\mathrm{m}^{-3}$ [20]. Electron temperatures T_e can be determined by measuring a spectral line emission ratio (eg. H-α/H-β). A comparison with theory then leads to $T_e < 1$ eV [21].

There exist many recombination and excitation processes from the ground state that can lead to the population of quantum states in the vicinity of $n \sim 30$. The list of reactions implemented in Yacora can be found in [22] or the above cited reference. We find that mainly recombination processes of ionic species determine the population density of highly excited Rydberg manifolds. The population of stronger bound levels, in contrast, gets quite rapidly dictated by collisional excitation from the ground state. This is a typical observation in such a so-called recombining plasma regime where $T_e < 1$ eV [23]. Extracting the exact contribution of each simulated channel to the Rydberg population coefficients remains however difficult due to the lack of knowledge of the involved ion densities and temperatures. Further diagnostic tools, like a Langmuir probe to precisely assess the electron temperature and most importantly the density of different ionic species in the discharge, would be needed to determine the relative importance of the different simulated excitation and recombination processes in the plasma.

We measured the distribution of the hydrogen quantum states emitted from the microwave discharge and detected a few hundred Hz increase in the count-rate of protons emerging from the atomic ionization process as a function of the electric field strength established between the ionizer meshes. The protons were detected in a single MCP configuration. The detector output was amplified and digitized. The events were discriminated as a function of the peak voltage of the few ten nanoseconds long charge burst detected on the anode of the MCP. The results and the potentials that were applied to the ionization meshes and the MCP are shown in Fig. 4 (top).

Each field ionizer setting probes a range of n-manifolds (for more details and formulas, please refer to chapter 5 of [4]). A graph of the proton detection rate as a function of the corresponding range of n (horizontal bars) is shown in Fig. 4 (bottom). Even though the data cannot yet be compared in a quantitative way to simulation results, qualitative features of the Rydberg state distribution can be extracted. The sudden drop for $n > 30$ can be explained by ionization most likely already within the few $10\,\mathrm{V\,cm^{-1}}$ stray field close to the microwave discharge region. The presence of states with $20 \leq n \leq 30$ is of high interest for deexcitation studies, especially because we expect the population of a large number of n-manifolds, which would reproduce conditions similar to those found in antihydrogen experiments. The scans are sensibly the same at room temperature and 25 K, where the blackbody irradiance in the frequency regime critical for ionization is reduced by more than an order of magnitude compared to 300 K. We conclude that the quantum states present in the beam are not very susceptible to being coupled to the continuum via THz radiation which is characteristic for high angular momentum states that exhibit minimum ionization cross sections (cf. Fig. 11 in [8]). In fact, one can indeed expect that non-circular states emerging from the plasma rather rapidly collapse to circular levels along their spontaneous decay cascade [24]. The lifetime τ of a (n, l) state with magnetic quantum number $|m| \leq l < n$ can be approximated by [25]

$$\tau = \left(\frac{n}{30}\right)^3 \left(\frac{l + 1/2}{30}\right)^2 \times 2.4\,\mathrm{ms}. \quad (1)$$

For beam velocities of the order of some $1000\,\mathrm{m\,s^{-1}}$ and the 0.7 m long flight path from the discharge plasma to the detection region, quantum states with lifetimes of the order of $100\,\mu\mathrm{s}$ can be expected to spontaneously decay before reaching the field ionization region (for $(n, l) = (16, 15)$ one finds $\tau \sim 97\,\mu\mathrm{s}$, cf. Eq. 1). Consequently, the signal levels off toward the low lying n states. At the same time, electron collisional processes start to play an increasingly important role toward stronger bound states which explains the intermittent rapid increase in rate between $15 < n < 20$.

For settings above $\sim 3.2\,\mathrm{kV}$, the cumulative detection rate drops from close to 400 Hz back to roughly 300 Hz at 3.5 kV. Part of an explanation for this might be the ionization of weakly bound atoms upstream the detection region by the increasingly large stray field emerging from the ionization stage. These protons must be expected to be less efficiently detected which can lead to a net drop in countrate on the MCP at high ionization field strengths under the assumption that no additional strongly bound states are accessed through the voltage ramp-up. Consequently, the region around and beyond $\sim 3.2\,\mathrm{kV}$ constitutes the maximal field ionization strength that can be employed and data at such high voltage configurations must be handled with care since this process might build up gradually.

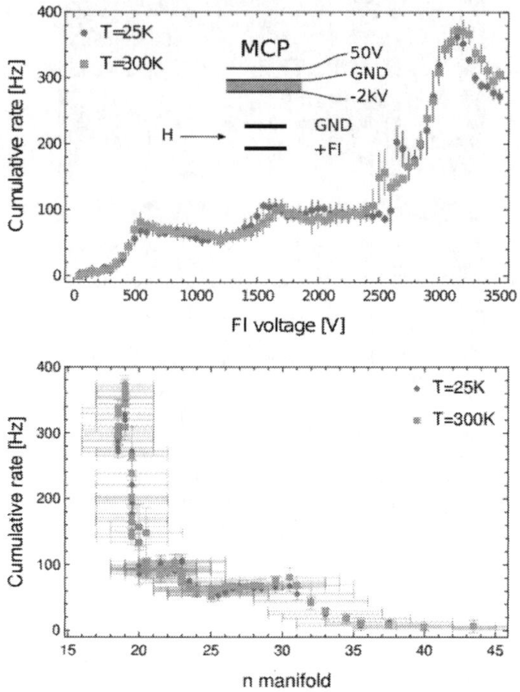

Figure 4: Top: Background subtracted rate of detected protons originating from the ionization of Rydberg hydrogen atoms as a function of the voltage applied to the field ionization meshes. Bottom: The same signal plotted as a function of the range of the atoms' principal quantum numbers n probed at each voltage setting (horizontal bars reflect the lack of knowledge of the distribution of substates within each n-manifold). We take into account a $\pm 0.5\,\mathrm{V}$ error on the potential applied to the field ionizer meshes (mounted 5 mm apart from each other) and a 10 % imprecision in the spacing between the potential grids. Both plots show results for two different temperatures of the cryogenic beam shield.

3 Summary and outlook

We reported on the design and on the status of a proof-of-principle experiment to demonstrate the performance of deexcitation techniques for antihydrogen atoms. An atomic hydrogen beam and an excitation laser have been developed to excite metastable 2s atoms toward Rydberg levels. An electron gun for the production of 2s states from the ground-state atoms emitted out of the microwave discharge plasma is currently being commissioned. We are in the process of designing a circular Rydberg production stage to introduce the required crossed electric and magnetic fields into the laser excitation region. This will allow for the production of single quantum states of interest for deexcitation studies.

In parallel, the production of a broad distribution of highly excited atoms with $20 \leq n \leq 30$ via recombination and collisional processes inside the discharge plasma has been experimentally evidenced relying on electric field ionization. This approach best reproduces the conditions faced in an antimatter experiment. We point out that this result can be of interest to the plasma community to, for example, benchmark collisional-radiative models in the high n-regime.

Suitable (anti)hydrogen mixing and deexcitation light sources have been tested in a cesium proof-of-principle experiment in [26] and in particular photomixing has been identified as a versatile and promising technology for our deexcitation purposes. The forthcoming step is the demonstration of fast stimulated Rydberg state mixing and deexcitation in hydrogen. We plan on developing a tailored photomixing device for the application in (anti)hydrogen experiments in the coming months.

The reported developments lay the foundation for a first stimulated deexcitation result in the near future. We then aim, in a subsequent step, for a swift installation of the commissioned and optimized technology in experiments at CERN's Antiproton Decelerator to enable the production of ground-state antihydrogen required to perform gravity and spectroscopy measurements in beam configurations.

Acknowledgements

This work has been sponsored by the Wolfgang Gentner Programme of the Bundesministerium für Bildung und Forschung (Grant No. 05E15CHA, university supervision by Norbert Pietralla) and the Austrian Science Fund (FWF): W1252-N27. It was supported by the Studienstiftung des Deutschen Volkes. In addition, the research presented was financed by the LASANTI grant (from Domaine d'Intérêt Majeur Science et Ingénierie en Région Île-de-France pour les Technologies Quantiques – DIM SIRTEQ –) and the Initiative de Recherche stratégique of the Université Paris-Saclay IQUPS (project number 156541). We warmly thank P. Chabert and C. Drag for useful discussions concerning radio-frequency plasma sources and the EP-AGS, EP-DI, EP-SAM, EN-AA and EP-ESE groups at CERN for the support provided for the laser room construction.

References

[1] C. Amsler, M. Antonello, A. Belov et al., Commun. Phys. **4**, 19 (2021)

[2] M. Ahmadi, B. Alves, C. Baker, W. Bertsche, A. Capra et al., Nat. Commun. **8**, 681 (2017)

[3] Y. Enomoto, N. Kuroda, K. Michishio, C.H. Kim, H. Higaki, Y. Nagata, Y. Kanai, H.A. Torii, M. Corradini, M. Leali et al., Phys. Rev. Lett. **105**, 24, 243401 (2010)

[4] B. Kolbinger, C. Amsler, S.A. Cuendis et al., Eur. Phys. J. D **75**, 91 (2021)

[5] E.D. Hunter, *Minimizing plasma temperature for antimatter mixing experiments* (2022), arXiv:2201.01256

[6] D. Comparat, C. Malbrunot, Phys. Rev. A **99**, 1, 013418 (2019)

[7] D. Comparat, C. Malbrunot, Phys. Rev. A **101**, 1, 019904 (2020)

[8] T. Wolz, C. Malbrunot, M. Vieille-Grosjean, D. Comparat, Phys. Rev. A **101**, 043412 (2020)

[9] C. Malbrunot, M. Diermaier, M. Simon et al., Nucl. Instrum. Methods Phys. Res. A **935**, 110 (2019)

[10] F.C. Fehsenfeld, K.M. Evenson, H.P. Broida, Rev. Sci. Instrum. **36**, 3, 294 (1965)

[11] P. Lottigier, A. Jucha, L. Cabaret et al., Appl. Phys. B **125**, 14 (2019)

[12] M.A. Luda, M. Drechsler, C.T. Schmiegelow, J. Codnia, Rev. Sci. Instrum. **90**, 2, 023106 (2019)

[13] T. Wolz, Ph.D. thesis, Technische Universität, Darmstadt (2022), in press

[14] A. Morgan, V. Zhelyazkova, S.D. Hogan, Physical Review A **98**, 4, 043416 (2018)

[15] R. Lutwak, J. Holley, P.P. Chang, S. Paine, D. Kleppner, T. Ducas, Phys. Rev. A **56**, 2, 1443 (1997)

[16] K.C. Harvey, J. Appl. Phys. **53**, 5, 3383 (1982)

[17] F. Biraben, J.C. Garreau, L. Julien, M. Allegrini, Rev. Sci. Instrum. **61**, 5, 1468 (1990)

[18] R.F. Stebbings, W.L. Fite, D.G. Hummer, R.T. Brackmann, Phys. Rev. **119**, 6, 1939 (1960)

[19] D. Wünderlich, M. Giacomin, R. Ritz, U. Fantz, J. Quant. Spectrosc. Radiat. **240**, 106695 (2020)

[20] P. Chabert, N. Braithwaite, *Physics of Radio-Frequency Plasmas* (Cambridge University Press, 2011)

[21] P.N. Kulkarni, *Rydberg Excitation of Hydrogen for studies towards Antihydrogen Deexcitation* (2021), Bachelor Thesis, Universität Leipzig

[22] *Yacora on the Web*, www.yacora.de, accessed: 2022-01-27

[23] M. Giacomin, *Application of collisional radiative models for atomic and molecular hydrogen to a negative ion source for fusion* (2017), Master Thesis, University of Padova

[24] M.R. Flannery, D. Vrinceanu, Phys. Rev. A **68**, 3, 030502 (2003)

[25] E.S. Chang, Phys. Rev. A **31**, 1, 495 (1985)

[26] M. Vieille-Grosjean, Z. Mazzotta, D. Comparat, E. Dimova, T. Wolz, C. Malbrunot, Eur. Phys. J. D **75**, 27 (2021)

Status of the Fermilab Muon $g-2$ Experiment

Paolo Girotti[1,2,*] *on behalf of the* Muon $g-2$ (E989) Collaboration

[1]University of Pisa
[2]INFN Pisa

Abstract. The muon anomalous magnetic moment, $a_\mu = (g-2)/2$, is a low-energy observable which can be both measured and computed with very high precision, making it an excellent test of the Standard Model (SM) and a sensitive probe of new physics. Recent efforts improved the precision of both the theoretical prediction and the experimental measurement. On the theory side, the Muon $g-2$ Theory Initiative, an international team of more than 130 physicists, reached in 2020 a consensus on the SM prediction for a_μ. On the experimental side, the E989 Muon $g-2$ Collaboration at Fermilab (FNAL) published in April 2021 a new measurement of a_μ from the Run-1 dataset (2018) with 0.46 ppm precision, corroborating the previous Brookhaven National Laboratory (BNL) measurement and increasing the discrepancy with the SM value to 4.2 standard deviations. In this paper, I will discuss the experimental setup and report on the current status of the experiment.

1 Introduction

The study of the lepton magnetic moments has a long history starting back in 1928 when Dirac predicted $g = 2$ for the electron. The 1948 calculation by J. Schwinger of the first-order correction due to virtual particles marked the beginning of a long sequence of improvements in both the theoretical prediction and the experimental measurement of the electron and muon anomalies. While the electron $g-2$ measurements represent a striking confirmation of QED theory, the muon anomaly has emerged as an important test of the entire Standard Model due to significant contributions from weak and strong interactions.
In April 2021, the Muon $g-2$ Collaboration released their Run-1 measurement of the muon anomaly with a precision of 460 ppb [1]. This result is in agreement with the previous experiment at Brookhaven National Laboratory (E821), which determined a_μ with an accuracy of 540 ppb [2]. By combining the two measurements into a new world average, we find a discrepancy of 4.2 σ with respect to the current consensus of the Standard Model prediction [3]. The current theoretical and experimental values are:

$$a_\mu^{Th}\ [2020] = 116\,591\,810(43)\times 10^{-11}\ (0.37\ \text{ppm}) \quad (1)$$

$$a_\mu^{Exp}\ [2021] = 116\,592\,061(41)\times 10^{-11}\ (0.35\ \text{ppm}) \quad (2)$$

$$a_\mu^{Exp} - a_\mu^{Th} = (251 \pm 59)\times 10^{-11}\ (4.2\sigma) \quad (3)$$

The history of $g-2$ is far from being completed, as the Muon $g-2$ Experiment at Fermilab will improve its precision by a factor of four, other experimental efforts are in the works, and new calculations with Lattice QCD techniques represent a new way to improve on the theoretical predictions.

*e-mail: pgirotti@fnal.gov

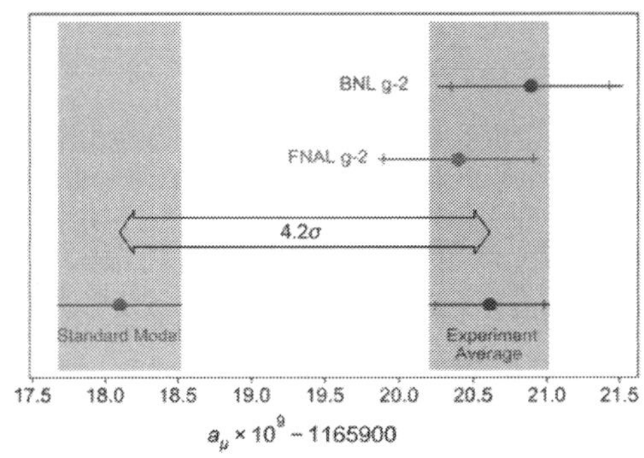

Figure 1: Run-1 FNAL measurement compared with the previous BNL experiment and the 2020 Standard Model consensus recommended by The Muon $g-2$ Theory Initiative [3]. Figure from [1].

1.1 The importance of g-2

The motivation for muon $g-2$ experiments comes from various reasons. First of all, the muon anomaly $a_\mu \equiv \frac{g-2}{2}$ is a relatively simple observable to measure. The g factor relates the magnetic moment of the muon with its spin:

$$\vec{\mu} = g\frac{q}{2m}\vec{S}. \quad (4)$$

Such measurements can be obtained through a spin-polarized beam of muons, an external and measurable magnetic field, and a technique to determine the spin precession through time. While such a setup is easy to create with current technology, many challenges arise when trying to achieve an accuracy of less than 1 part-per-million (ppm).

Despite the difficulties, we pursue precision measurements of a_μ because the anomaly probes all of the interactions between the lepton and virtual particles. The whole Standard Model is tested, since all possible particles can contribute either via direct virtual interaction with the muons or with higher-order loop corrections, adding to the value of a_μ. Figure 2 depicts the lower-order interactions for the electromagnetic, weak, and strong forces. Table 1 lists the contributions to a_μ by the three SM forces.

Moreover, eventual measurements that differ from the SM

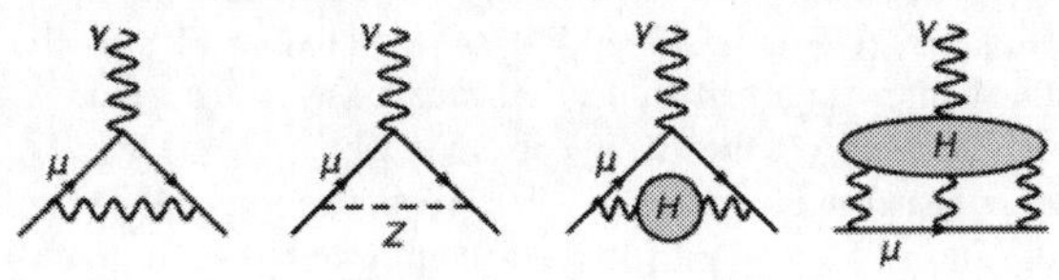

Figure 2: Lower-order examples for EM, weak, and QCD interactions [1].

theoretical prediction would indicate hints of *new physics*. Such results would help setting new constraints on the multitude of currently proposed Beyond Standard Model (BSM) theories.

Muons are particularly important in such a measurement because BSM interactions with massive particles contribute with mass suppression terms, $\propto (\frac{m_{lepton}}{M})^2$. The electron $g-2$ has already been measured with around $O(10^3)$ times more precision than the muon [4], but the relative mass ration between the electron and the muon enhances the sensitivity to these terms by a factor of $(\frac{105.66}{0.511})^2 \approx 43000$ [5]. Following the same reasoning, τ leptons would be even better for detecting new heavy particles, but their very short lifetime (2.9×10^{-13} s), multiple decay branches, and other practical reasons make such an experiment not feasible with current technology.

Table 1: Standard model contributions to a_μ [3].

Term	Value $(\cdot 10^{-11})$
a_μ^{QED}	$116\,584\,718.931 \pm 0.104$
a_μ^{EW}	153.6 ± 1.0
a_μ^{HVP}	$6\,845 \pm 40$
a_μ^{HLBL}	92 ± 18
Total SM	$116\,591\,810 \pm 43$

2 The experiment

The new Muon $g-2$ Experiment in operation at Fermi National Accelerator Laboratory aims to measure the muon's anomalous magnetic moment with a precision of 0.14 ppm, a factor of four better than the previous BNL E821 Experiment. To achieve this precision, a statistics of ~20 times the amount collected at BNL is needed, and the systematics must be contained within 100 ppb. In order to obtain a large number of observed muons, the BNL storage ring was moved from Brookhaven to Fermilab and installed in the FNAL muon campus accelerator chain, where the beam is cleaner and much more intense.

The experimental technique consists of producing a polarized and clean beam of muons, sending it to a storage ring with very uniform magnetic field, and observing the decay positrons. Three measurements are needed to calculate the muon anomaly a_μ: the muon anomalous precession frequency, the magnetic field intensity, and the beam distribution inside the storage region. The master formula is the following:

$$a_\mu = \frac{\omega_a}{\tilde{\omega}_p} \frac{g_e}{2} \frac{m_\mu}{m_e} \frac{\mu_p}{\mu_e}, \quad (5)$$

where ω_a is the muon anomalous precession frequency, and $\tilde{\omega}_p$ is the Larmor precession frequency of the proton (ω_p) convoluted with the beam distribution, representing the average field intensity experienced by the muons. The remaining factors are known with sufficient precision from other experiments.

2.1 The muon beam

The muons observed in the Muon $g-2$ Experiment originate from decaying pions, which are in turn produced by the collisions of an 8 GeV proton beam on an *Inconel®* 600 target. After the collisions, secondary particles are focused with an electrostatic lithium lens, and pions having 3.1 GeV momentum are extracted. The beam of pions then circulates inside a Delivery Ring, where remaining protons get removed and all the pions decay into muons. Since pions have zero spin, the muons are emitted isotropically in the rest frame of reference, but their helicity is constrained by the weak decay, as illustrated in Fig. 3. By selecting boosted muons with higher momentum, it is possible to obtain a highly polarized beam in the laboratory frame of reference. Finally, the 120-ns

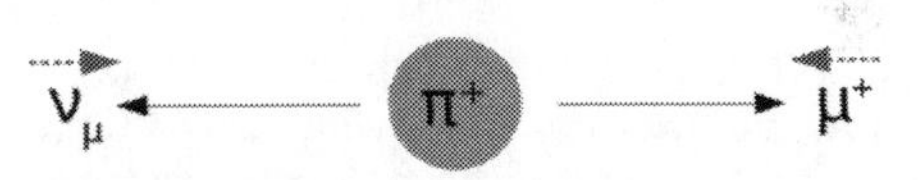

Figure 3: Decay of the positive pion in the rest frame of reference. The spin of the muon points toward the pion as the neutrino must be left-handed because of parity violation of the weak interaction.

long muon bunches enter the Muon $g-2$ storage ring at an instantaneous rate of 100 Hz (11.4 Hz average). A superconducting inflector magnet lets the beam pass through the ring yoke into the storage region. Then, after the first quarter of an orbit, a fast pulsed kicker magnet deflects the muon bunch into the final orbit, which has a radius of 7.112 m and a period of 149.2 ns. The ~5000 stored muons per bunch then circulate inside the ring for 700 μs, while a set of electrostatic quadrupoles provides weak focusing for vertical confinement.

2.2 The field measurement

The magnet field measurement consists of various components and techniques used to determine the value of ω_p:

378 Nuclear Magnetic Resonance (NMR) fixed probes are placed along the ring under and over the vacuum chamber. The probes keep monitoring the field during the whole data taking. Every three days during data taking, trolley runs are performed in which a mobile cylinder equipped with 17 NMR probes traverses the beam storage region to make 9000 measurements along the ring [6]. The trolley measurements produce a three-dimensional (3D) map of the magnetic field inside the storage region. The 3D map is then interpolated over the time between two consecutive trolley runs with the fixed probes data. While the magnetic field is ppm-level uniform along the beam cross section, it is necessary to know the muon distribution inside the storage ring in order to achieve an uncertainty of 70 ppb on $\tilde{\omega}_p$. For this, a set of two tracker stations observe positron tracks and extrapolate the decay vertex in the storage orbit. The beam distribution is extrapolated over the rest of the ring with data from the calorimeters and multiple dedicated simulations [7].

2.3 Muon precession frequency measurement

A charged particle with mass m placed in an uniform external magnetic field will follow a circular path because of the the Lorentz force, and this motion is called cyclotron motion. If the particle has spin, the spin direction will also rotate (precess) around the direction of the magnetic field. In the absence of electrical fields, and with the particle velocity and spin perpendicular to the magnetic field, the equations governing this motion are:

$$\vec{\omega}_s = -\frac{ge\vec{B}}{2m} - (1-\gamma)\frac{e\vec{B}}{m\gamma} \tag{6}$$

$$\vec{\omega}_c = -\frac{e\vec{B}}{m\gamma}\,, \tag{7}$$

where ω_s is the spin precession angular frequency and ω_c is the cyclotron frequency. The second term of Eq. 6 is the Thomas correction to the Lorentz force, which accounts for the rotation of the particle's frame of reference. Subtracting Eq. 7 from Eq. 6, we obtain the rotation of the particle spin with respect to its momentum. This is the anomalous precession frequency ω_a:

$$\vec{\omega}_a \equiv \vec{\omega}_s - \vec{\omega}_c \simeq -\left(\frac{g-2}{2}\right)\frac{e\vec{B}}{m} \equiv -a_\mu\frac{e\vec{B}}{m}\,. \tag{8}$$

However, the particle beam is not always perfectly parallel to the storage plane, but oscillates and breathes vertically and horizontally. The anomalous precession frequency is sensible to such effects:

$$\vec{\omega}_a = -\frac{e}{m}\left[a_\mu\vec{B} - \left(a_\mu - \frac{1}{\gamma^2-1}\right)\frac{\vec{\beta}\times\vec{E}}{c} - a_\mu\left(\frac{\gamma}{\gamma+1}\right)(\vec{\beta}\cdot\vec{B})\vec{\beta}\right]. \tag{9}$$

The storage region houses several electrostatic quadrupoles for a weak vertical focusing of the beam. This would affect the muon precession frequency, but, for a muon beam with *magic* momentum $p_\mu = 3.094$ GeV/c, corresponding to a value of $\gamma = 29.3$, the second term of Eq. 9 cancels out. The remaining effect and the third term are measured separately and applied as *E-field* and *Pitch* corrections to ω_a.

The measurement principle relies on the parity-violating nature of the weak decay. The positive muons decay into a positron and two neutrinos with nearly 100% probability. In the rest frame of the muon, the highest energy decay positrons come from decays in which the neutrinos are emitted collaterally, as depicted in Fig. 4. In this scenario, half of the initial rest mass of the muon is carried away by the decay positron ($E_{max} \approx 53$ MeV), while the rest is shared by the two neutrinos. Since the neutrino and anti-neutrino are traveling in the same direction, and the weak decay dictates they must have opposite helicities, their spins must be opposite. With the neutrinos' spins canceling, conservation of angular momentum forces the decay positron to carry the spin of the parent muon. The $V-A$ nature of the weak decay prefers to couple to a right-handed positron, so the high-energy decay positron depicted in Fig. 4 tends to be emitted in the direction of the muon spin. Therefore, in the rest frame of the muon, the spin direction of the muon can be monitored by observing the direction at which the high energy decay positrons are emitted. The Muon $g-2$ Experiment is equipped with 24 electromagnetic calorimeters located around the ring with 15° azimuthal distance between each other. The asymmetry of the decay process, together with the fact that the spin precesses with respect to the momentum, results in an oscillation in the count of positrons over time. The number of detected positrons above a single energy threshold E_{th} is:

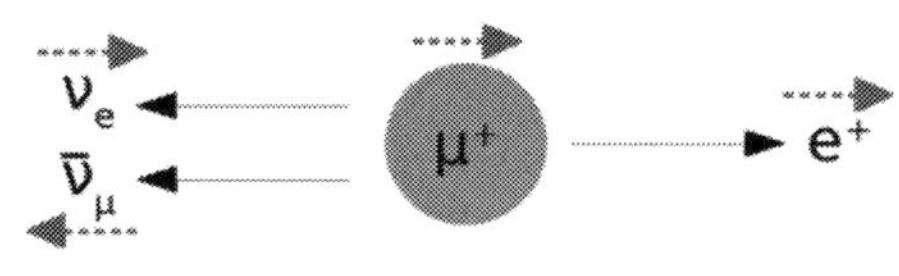

Figure 4: Muon decay configuration that maximizes the positron momentum and decay probability. The right-handed positron is emitted in the direction of the muon spin.

$$N(t) = N_0 e^{-t/\tau}[1 + A\cos(\omega_a t + \phi)]\,, \tag{10}$$

where the normalization N_0, the asymmetry A and the initial phase ϕ are all dependent on the energy threshold. τ represents the lifetime of the muon in the laboratory frame of reference, that is $\gamma\tau_\mu \simeq 64.4$ μs. Additional beam-related effects, as coherent betatron oscillations and muon losses, appear as multiplicative terms to N_0, A, and ϕ. Three different reconstruction techniques and several analysis procedures are implemented to obtain the ω_a value [8]. As an example, in the A-Weighted method the positrons counts are weighted by the asymmetry, which depends on the energy, yielding the maximum possible statistical power for a given threshold E_{th}. The typical plot showing the count of positrons over time is called *wiggle plot* and is shown in Fig. 5.

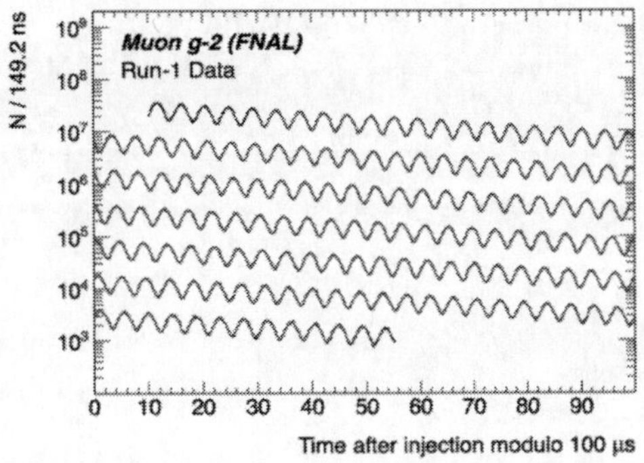

Figure 5: *Wiggle plot* showing the number of >1.7 GeV positrons over time. The graph is wrapped every 100 μs to show the entire measurement period. The oscillation period is $2\pi/\omega_a$.

The positrons are measured by 24 calorimeters, which each have a 9x6 array of PbF_2 crystals with size $2.5 \times 2.5 \times 14$ cm. The charged particles in the electromagnetic shower generate Čerenkov photons. The choice of a pure Čerenkov material is driven by the almost instantaneous signal produced when a positron strikes, further enhanced by a black Tedlar® wrapping that absorbs part of the reflected photons. Every crystal is coupled with a Silicon PhotoMultiplier (SiPM) detector, whose signals are digitized at an 800 MHz sampling rate and chopped into 40-ns islands by on-line GPUs. A precise gain calibration of the SiPMs at the level of $O(10^{-4})$ is provided by a Laser-based Calibration System [9].

2.4 Blinding

The Muon $g-2$ Experiment has an unique feature, which is the hardware blinding of the digitization clocks. Before the start of each run of data acquisition, the main clock that determines the calorimeters digitization rate is configured with a secret offset in the 25 ppm range. The real frequency is known only by two persons outside of the collaboration, and that number is only revealed to the analyzers at the final stage of the measurement process. The whole muon precession analysis is therefore performed blindly on time-uncalibrated data. When the collaboration unanimously decides that the analysis is mature and the final value is calculated, the secret frequency is revealed.

3 Status and future of the Muon $g-2$ Experiment

The Muon $g-2$ Experiment at Fermilab is currently running its fifth year of data taking. In the first four years, an equivalent statistics of 13×BNL has been collected (Fig. 6). The Run-1 data has been completely analyzed and has been published earlier this year. The result was in agreement with the BNL experiment and the combined value increased the discrepancy with theory to 4.2 σ [1].

The data from Run-2 through Run-4 data is currently under reconstruction and analysis, with many efforts in improving the systematic errors with respect to Run-1, both on the hardware and the software fronts. For instance, the magnet has been covered with a thermal blanket, and the hall is now temperature-controlled with air conditioners. The kicker strength has been improved [10], and the electrostatic quadrupole resistors have been upgraded. New beam scraping techniques and quadrupole radio-frequency pulses have been put in place to increase the storage efficiency and minimize the betatron oscillations [11]. On the analysis side, independent efforts from the ω_a groups try to increase pileup separation and decrease gain related systematics. The beam-related effects to ω_a are being calculated with improved precision thanks to ever increasing simulation studies. The Run-5 data taking just started in late 2021, and another Run-6 is currently planned. The collaboration is considering the possibility of using negative muons for the Run-6 data taking. Such a run would provide an opportunity to test the robustness of the method and verify CPT and Lorentz conservation.

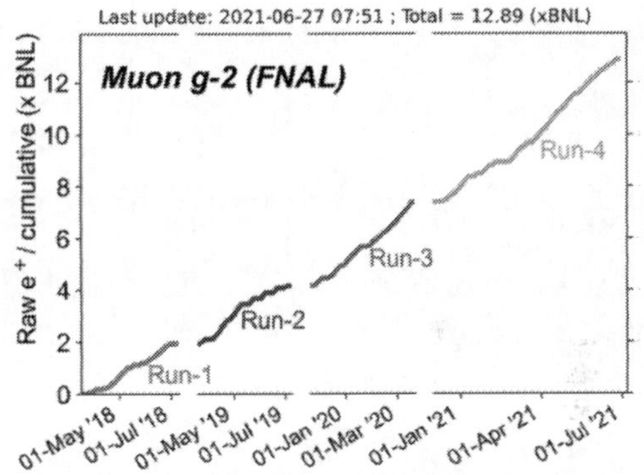

Figure 6: Collected statistics compared to the BNL dataset in the first four runs of the experiment.

4 Acknowledgments

This work was supported in part by the US DOE, Fermilab, the Istituto Nazionale di Fisica Nucleare (Italy), and the European Union's Horizon 2020 research and innovation programme under the Marie Skłodowska-Curie grant agreements No. 690835 (MUSE) and No. 734303 (NEWS).

References

[1] B. Abi *et al.* (Muon $g-2$ Collaboration), Phys. Rev. Lett. **126**, 141801 (2021)

[2] G. W. Bennett *et al.* (Muon $g-2$ Collaboration), Phys. Rev. D **73**, 072003 (2006)

[3] T. Aoyama *et al.*, Physics Reports **887** (2020)
[4] D. Hanneke, S. Fogwell Hoogerheide, and G. Gabrielse., Phys. Rev. A **83**, 052122 (2011)
[5] F. Jegerlehner and A. Nyffeler, Phys. Rept. **477**, pages 1-110 (2009)
[6] T. Albahri *et al.* (Muon $g-2$ Collaboration), Phys. Rev. A **103**, 042208 (2021)
[7] T. Albahri *et al.* (Muon $g-2$ Collaboration), Phys. Rev. Accel. Beams **24**, 044002 (2021)
[8] T. Albahri *et al.* (Muon $g-2$ Collaboration), Phys. Rev. D **103**, 072002 (2021)
[9] A. Anastasi *et al.*, JINST **14**, P11025 (2019)
[10] A.P. Schreckenberger *et al.*, Nucl. Instrum. Meth. A **1011**, 165597 (2021)
[11] On Kim *et al.*, New J. Phys. **22**, 063002 (2020)

Recent progress of laser spectroscopy measurements of pionic helium

Masaki Hori[1,2,*], *Hossein* Aghai-Khozani[1,**], *Anna* Sótér[1,***], *Andreas* Dax[3], and *Dániel* Barna[4,****]

[1]Max-Planck-Institut für Quantenoptik, Hans-Kopfermann-Strasse 1, D-85748 Garching, Germany
[2]Fakultät für Physik, Ludwig-Maximilians-Universität München, 80799 Munich, Germany
[3]Paul Scherrer Institut, Forschungsstrasse 111, CH-5232 Villigen, Switzerland
[4]CERN CH-1211, Geneva 23, Switzerland

Abstract. We review the results of recent laser spectroscopy experiments on metastable pionic helium atoms at the Paul Scherrer Institute's 590 MeV cyclotron facility that was carried out by the PiHe collaboration. Some future perspectives are briefly discussed.

1 Introduction

Metastable pionic helium atoms ($\pi^4\text{He}^+ \equiv \pi^- + {}^4\text{He}^{2+} + e^-$) [1–5] consist of a helium nucleus, an electron in the ground state, and a negatively-charged pion in a state with high principal and orbital angular momentum quantum numbers $n \approx \ell + 1 \approx 16$. Atoms populating such states possess nanosecond-scale lifetimes against π^- nuclear absorption because the pionic orbitals have very small overlap with the helium nucleus. The deexcitation rates of the atom via electromagnetic cascade processes such as Auger and radiative decays are thus highly reduced. The long lifetime recently enabled laser spectroscopy [6, 7] of an infrared transition of $\pi^4\text{He}^+$ to be carried out. This constituted the first laser spectroscopy measurement of a mesonic atom, and showed the existence of this three-body variant of pionic helium. The π^- mass [8–10] can in principle be determined with a high precision by comparing the atomic transition frequencies with the results of quantum electrodynamics (QED) calculations, as has been done for three-body antiprotonic helium ($\overline{p}\text{He}^+ \equiv \overline{p} + \text{He}^{2+} + e^-$) atoms [11–24]. The π^- mass can then be used to help set improved upper limits on constraints on the muon antineutrino mass by laboratory experiments [25]. Some upper limits may also be set on any exotic force [26–31] that involves the π^-. The atomic structure of $\pi^4\text{He}^+$ has no hyperfine structure that arises from the spin-spin interaction between the spin-0 π^- and ^{4}He nucleus [32, 33], and so is interesting from the point of view of studying the bound-state QED of spin-0 systems. Several past experiments [34–38] based on either bubble chambers or particle counters have indirectly inferred the existence of $\pi^4\text{He}^+$ atoms by observing that some π^- coming to rest in helium targets have an anomalously long lifetime. Quantitative comparisons

*e-mail: Masaki.Hori@cern.ch
**Present address: McKinsey and Company, Sophienstrasse 26, 80333 Munich, Germany
***Present address: ETH Zürich, IPA, Otto-Stern-Weg 5, 8093 Zurich, Switzerland
****Present address: Institute for Particle and Nuclear Physics, Wigner Research Centre for Physics, H-1525 Budapest 114, P.O.B. 49, Hungary

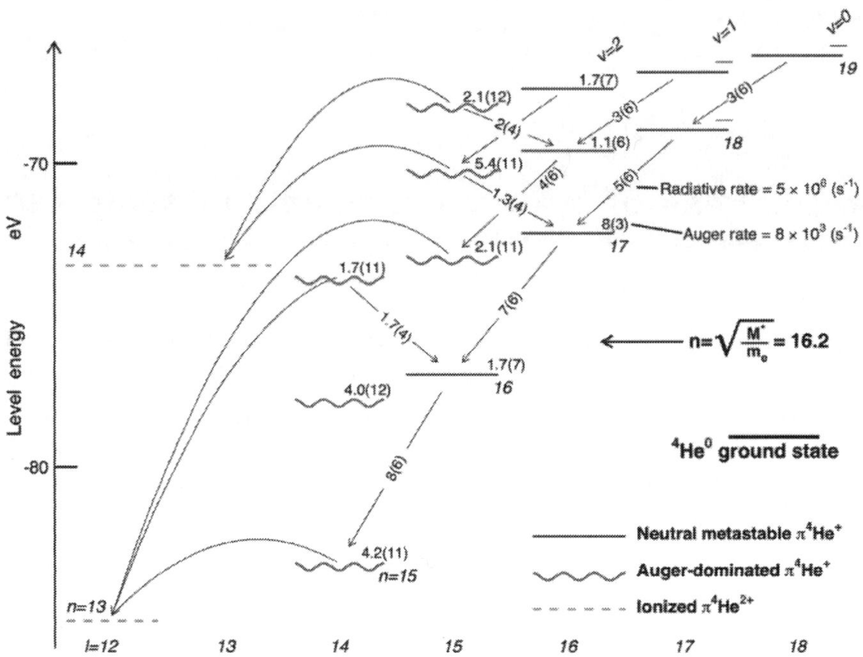

Figure 1. Energy level diagram of $\pi^4\text{He}^+$. The theoretical absolute energies of the states (n, ℓ) are plotted relative to the three-body-breakup threshold. The wavy lines indicate Auger-dominated states with picosecond-scale lifetimes, the solid lines metastable states with > 10 ns lifetimes. The Auger decay rates of each state are shown in s^{-1}. The dashed lines correspond to $\pi^4\text{He}^{2+}$ ionic states that are formed after Auger emission. The curved arrows show the Auger transitions that have minimum $|\Delta\ell_A|$. The radiative transitions $(n, \ell) \to (n-1, \ell-1)$ and $(n, \ell) \to (n-1, \ell+1)$ are indicated by straight arrows, with the corresponding decay rates in s^{-1}. From Ref. [7].

of the experimental data with the theoretical calculations have been difficult, however, as some sets of calculated $\pi^4\text{He}^+$ decay rates have differed from each other by 1–2 orders of magnitude [2, 7, 39].

2 Experimental method

The recent experiment utilized the πE5 beamline of the Paul Scherrer Institute's 590 MeV ring cyclotron facility that provided a π^- beam that had a momentum between $P = 83$ and 87 MeV/c, and an average intensity of $N_\pi = (2-3) \times 10^7\ \text{s}^{-1}$. Infrared sub-nanosecond laser pulses excited a transition from a pionic state $(n, \ell) = (17, 16)$ of the neutral atom that had a nanosecond-scale lifetime, to another state $(17, 15)$ that had a lifetime of $\tau \approx 5$ ps against Auger decay [7] (see energy level diagram of Fig. 1). A two-body pionic helium ion ($\pi^4\text{He}^{2+} \equiv \pi^- + \text{He}^{2+}$) was formed after Auger emission of the remaining 1s electron. Atomic collisions in the experimental target caused Stark mixing between the Rydberg ionic states and the states with low ℓ values [40] as well as other possible cascade effects [41] that eventually lead to the absorption of the π^- by the helium nucleus. The high-precision laser spectroscopic study of this ion would also be interesting [42]. This allowed the resonance condition between the laser and the $\pi^4\text{He}^+$ atom to be detected as a sharp peak in the rates of neutrons, protons, and deuterons that emerged from the nuclear absorption. The experiment involved numerous backgrounds of much higher intensity than the laser resonance signal.

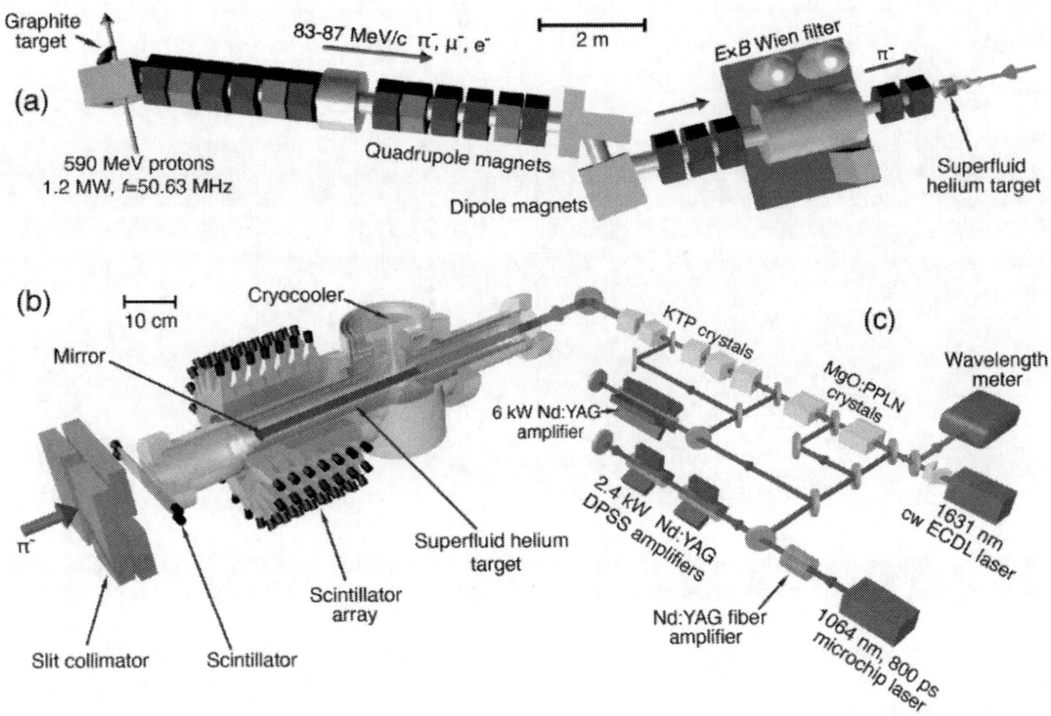

Figure 2. (a): Schematic layout of the πE5 experimental beamline showing the positions of the Wien filter and superfluid helium target. (b): Layout of the experimental target. (c) Schematic of the laser system, see text. From Ref. [6].

The measurement therefore required detectors of 1 ns-scale timing resolution and moderate energy resolution to isolate the signal.

A Wien filter was placed in the beamline upstream of the experiment and used to divert most of the contaminant e^- that arrived at a rate $> 3 \times 10^9$ s^{-1} into the blades of a collimator. The remaining π^- beam was focused into a superfluid helium target. The beam first passed through a segmented plastic scintillator plate with a thickness $t_d = 4.7$ mm. The π^- arrived at the experimental target at timings spaced by regular intervals $\Delta t = 19.75$ ns, which arose from the $f_a = 50.63$ MHz acceleration radiofrequency (RF) of the cyclotron. Each RF cycle contained on average $N_\pi/f_a \approx 0.4 - 0.6$ pions. A $\approx 2\%$ fraction of the π^- that came to rest in the helium target (see Fig. 2 (a)) with a length of 150 mm, diameter of 42 mm, and a temperature of $T = 1.7$ K formed metastable π^4He^+. A 800 ps-long laser pulse of diameter $d = 25$ mm, pulse energy $E = 10$ mJ, repetition rate $f_r = 80.1$ Hz and wavelength $\lambda \approx 1631$ nm entered the target through a pair of fused silica windows and irradiated the atoms. The estimated π^4He^+ production rate of $> 3 \times 10^5$ s^{-1} implied a probability of coincidence of 10^{-3} for a laser pulse to irradiate a long-lived atom.

The nuclear fragments that emerged from the π^- absorption [7] with a typical kinetic energy of a few tens of MeV were measured by an array of 140 plastic scintillation counters that covered a solid angle of $\approx 2\pi$ steradians seen from the target. The size $40 \times 35 \times 34$ mm^3 of each counter was chosen to obtain a significant ($< 10\%$) detection efficiency for $E \geq 25$ MeV neutrons [7], while allowing the rejection of most of the background e^- that were produced either from μ^- decay or the particle beam. These e^- deposited an average

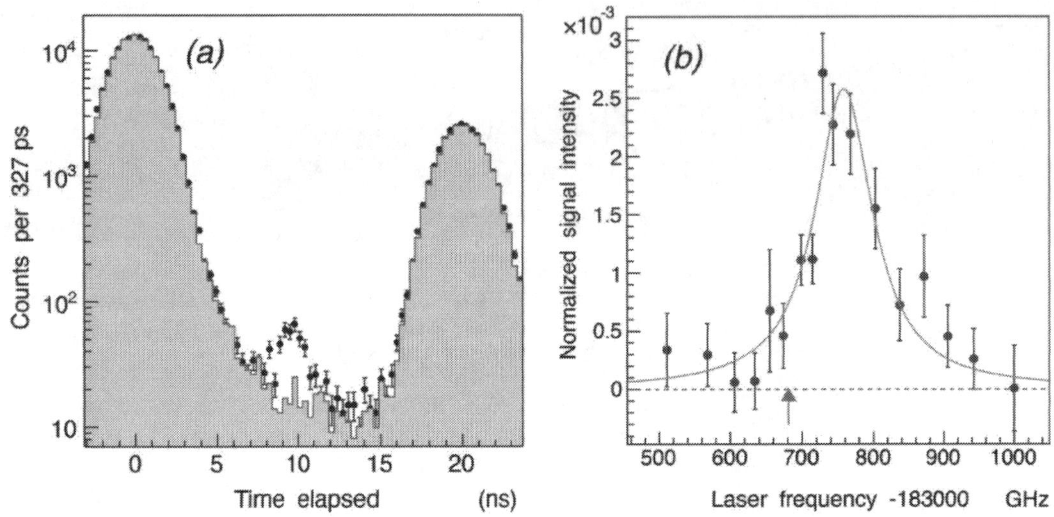

Figure 3. (a): Time spectra of nuclear fragments (mainly protons, neutrons, and deuterons) that emerged from π^- absorption in helium nuclei measured with (indicated by filled circles) and without (blue histogram) the laser irradiation of the target. The peak in the former spectrum at $t = 9$ ns shows the laser resonance signal of the transition $(n, \ell) = (17, 16) \to (17, 15)$. (b): The spectral profile of the transition as a function of the laser frequency. The red arrow indicates the position of the theoretical transition frequency, see text. From Ref. [6].

energy $\Delta E = 6 - 8$ MeV in the counters. We developed a data acquisition system based on waveform digitizers to record the signal waveforms of the 140 counters during each laser pulse arrival [43–45] at a sampling rate $f = 3.06$ Gs·s^{-1}. An earlier version of the electronics was used by the ASACUSA collaboration to determine upper limits on the cross sections of antiprotons with kinetic energy $E \approx 125$ keV annihilating in thin target foils [44, 46, 47]. The results of the experiment were compared with the annihilation cross sections measured at antiproton energies $E = 5.3$ MeV [48, 49].

The blue histogram of Fig. 3 (a) represents the time distribution of scintillator hits that were measured without laser irradiation. The π^- arrivals at $t = 0$ and 19.75 ns produced peaks in the spectrum that contained > 97% of the π^- that were immediately absorbed by the helium nuclei. The metastable atoms that decayed spontaneously with a lifetime of $\tau = (7 \pm 2)$ ns were contained in the remaining (2.1 ± 0.7)% fraction of the events in the intervals between the π^- arrivals. The shape of this experimental spectrum is in rough agreement with the results of a Monte Carlo simulation [7].

The laser pulses were produced by an injection-seeded, optical parameteric generator (indicated as OPG in Fig. 2(b)) and amplifier (OPA). The laser was precisely fired in synchronization with the RF signal of the cyclotron by pumping the OPG-OPA with a diode-pumped solid state (DPSS) neodymium-doped yttrium aluminium garnet (Nd:YAG) laser of single pass design. The seed light for the OPG-OPA was produced by a continuous-wave (cw) external-cavity diode laser (ECDL). The OPG was carried out using magnesium oxide doped periodically-polled lithium niobate (MgO:PPLN) crystals which produced laser pulses of energy $E = 70$ uJ. The beam was then amplified via OPA to $E = 10$ mJ using five potassium titanyl phosphate (KTP) crystals. The portion of the laser beam with a narrow spectral component had a linewidth of around 10 GHz. The OPG and OPA processes caused a 3 GHz uncertainty in the determination of the laser optical frequency.

3 Experimental results

We first searched for the $(n, l) = (16, 15) \rightarrow (17, 14)$ transition by irradiating the target with resonant laser pulses that were generated by a laser based on dye and Ti:sapphire [50] pulsed amplification. The laser was scanned over a 200 GHz range around the theoretical transition frequency, but no resonance was detected. Theoretical calculations [3] show that the resonance daughter state $(n, \ell) = (17, 14)$ strongly couples to an electronically excited state of π^4He^+. This is expected to destabilize the daughter state during atomic collisions, as was seen in the case of $\overline{p}He^+$ [51–53].

We next searched for the resonance $(16, 15) \rightarrow (16, 14)$ by tuning the laser to the theoretical transition wavelength $\lambda = 1515.3$ nm. The daughter state $(16, 14)$ is predicted to have a lifetime $\tau = 250$ fs [7] which corresponds to a large resonance width $\Gamma_A = 640$ GHz. We accumulated $> 6 \times 10^7$ detected π^- arrivals over several weeks, but no statistically significant signal was detected. The reason for this non-observation is not understood, but collisions with other helium atoms may destroy the π^- population that occupies the resonance parent state $(16, 15)$. Atomic collisions have been observed to similarly shorten the lifetimes of some $\overline{p}He^+$ states [18, 19, 54–59]. Another possible reason for the non-observation may be that a negligible fraction of π^- are captured into the parent state $(16, 15)$. Some lower-n states of $\overline{p}He^+$ have shown a small primary population [56, 60–64].

We then searched for the transition $(17, 16) \rightarrow (17, 15)$. The time spectrum corresponding to 2.5×10^7 π^- arrivals that was measured with the wavelength of the OPG-OPA laser tuned to $\lambda \approx 1631.4$ nm is plotted in Fig. 3(a) using filled circles with error bars. A peak containing about 300 events was observed at $t \approx 9$ ns, with a signal-to-noise ratio of 4 and a statistical significance corresponding to > 7 standard deviations. The $\Delta t = 2$ ns width of the peak is compatible with the dispersion of the time-of-flights of the fission fragments that arrived at the scintillator array. The detected rate of resonant π^4He^+ events of 3 h^{-1} is roughly compatible with the production rate of $> 3 \times 10^5$ s^{-1} of the pionic atoms and with the above Monte Carlo simulations [7]. The simulation assumed that most of the metastable atoms populate the parent state $(17, 16)$. As expected, the signal decreased and disappeared when the laser was detuned off the resonance frequency.

The resonance profile shown in Fig. 3 (b) was obtained by scanning the laser frequency and plotting the number of detected events under the laser-induced peak at $t \approx 9$ ns. Each data point contains data collected over a 20–30 h period. The statistical uncertainty arising from the finite number of resonant π^4He^+ events is indicated by the vertical error bars. The measured resonance width of ≈ 100 GHz is believed to arise from the 33 GHz Auger width [7] of the daughter state $(17, 15)$ predicted by theory, collisional and power broadening effects [4] which are estimated to cause a contribution of ≈ 50 GHz, and the ≈ 10 GHz linewidth of the narrowband spectral component of the OPG-OPA laser. Atomic collisions that shorten [3, 54] the lifetime of the resonance daughter state may cause an additional broadening. The interaction between the electron spin and the orbital angular momentum of π^- is expected to give rise to a pair of fine structure sublines that are spaced by a frequency interval of 3.0 GHz [7, 19]. These sublines cannot be resolved in this spectrum, since the splitting is much smaller than the natural width of the resonance. The reduced χ^2 value of the best fit (indicated by the blue curve) of two overlapping Lorentzian functions that take these fine structure sublines into account was 1.0.

The resonance centroid was determined as $\nu_{exp} = 183760(6)(6)$ GHz. The statistical uncertainty due to the finite number of detected π^4He^+ events is 6 GHz, whereas the systematic uncertainty of 6 GHz contains the contributions related to the selection of the fit function and the determination of the laser frequency. The experimental frequency ν_{exp} is larger than the theoretical value [7] $\nu_{th} = (183681.8 \pm 0.5)$ GHz by $\Delta\nu = (78 \pm 8)$ GHz. This difference is believed to be caused by atomic collisions that shift the resonance frequency [4]. Similar shifts have previously been observed [20, 54, 58, 65] in some $\overline{p}He^+$ resonances. A theoretical calculation based on the impact approximation of the binary collision theory of spectral lineshapes predicted that the gradient of this shift at a target temperature $T = 4$ K is $d\nu/d\rho = (4.4 - 6.5) \times 10^{-21}$ GHz·cm^3 [4]. The blueshift at superfluid density expected from this calculation corresponds to between $\Delta\nu = 96$ and 142 GHz, which roughly agrees with the experimental result. This shift must be experimentally measured in future experiments before the π^- mass can be precisely determined.

We plan to search for other transitions such as $(17, 16) \rightarrow (16, 15)$ that are expected to be narrower by a factor of at least 10^{-3} compared to the resonance observed in the above experiment. Helium gas targets in which the collisional shifts are small may be used, together with various laser spectroscopic techniques [14, 17, 21, 50]. The precision of the theoretical calculation of the transition frequencies is currently limited by the experimental uncertainty of the π^- mass. The precision of the QED calculations themselves [7] may be improved to a fractional precision of better than 10^{-8} for some transitions, as in the cases of HD^+ [66, 67] molecular ions and $\overline{p}He^+$ [11, 12] atoms. These experiments at PSI will complement the $\overline{p}He^+$ experiments carried out by the ASACUSA collaboration at the ELENA facility [68–71]. In the far future, these laser spectroscopy experiments may lead to studies of the shifts of the atomic binding energies that arise from the strong interaction [72]. Some novel methods of producing pionic atoms at the proposed gamma factory at CERN [73] may also be studied.

References

[1] G.T. Condo, Phys. Lett. **9**, 65 (1964)
[2] J.E. Russell, Phys. Rev. Lett. **23**, 63 (1969)
[3] V.I. Korobov, A.K. Bekbaev, D.T. Aznabayev, S.A. Zhaugasheva, J. Phys. B **48**, 245006 (2015)
[4] B. Obreshkov, D. Bakalov, Phys. Rev. A **93**, 062505 (2016)
[5] D. Baye, J. Dohet-Eraly, Phys. Rev. A **103**, 022823 (2021)
[6] M. Hori, H. Aghai-Khozani, A. Sótér, A. Dax, D. Barna, Nature **581**, 37 (2020)
[7] M. Hori, A. Sótér, V.I. Korobov, Phys. Rev. A **89**, 042515 (2014)
[8] S. Lenz et al., Phys. Lett. B **416**, 50 (1998)
[9] M. Trassinelli et al., Phys. Lett. B **759**, 583 (2016)
[10] M. Daum, R. Frosch, P.-R. Kettle, Phys. Lett. B **796**, 11 (2019)
[11] V.I. Korobov, L. Hilico, J.-P. Karr, Phys. Rev. Lett. **112**, 103003 (2014)
[12] V. Korobov, L. Hilico, J.-P. Karr, Phys. Rev. A **89**, 032511 (2014)
[13] V.I. Korobov, Phys. Rev. A **89**, 014501 (2014)
[14] M. Hori et al., Phys. Rev. Lett. **96**, 243401 (2006)
[15] T. Pask et al., J. Phys. B **41**, 081008 (2008)
[16] S. Friedreich et al., Phys. Lett. B **700**, 1 (2011)
[17] M. Hori et al., Nature **475**, 484 (2011)
[18] T. Kobayashi et al., J. Phys. B **46**, 245004 (2013)
[19] S. Friedreich et al., J. Phys. B **46**, 125003 (2013)
[20] A. Adamczak, D. Bakalov, Phys. Rev. A **88**, 042505 (2013)
[21] M. Hori et al., Science **354**, 610 (2016)
[22] M.H. Hu, S.M. Yao, Y. Wang, W. Li, Y.Y. Gu, Z.X. Zhong, Chem. Phys. Lett. **654**, 114 (2016)
[23] D. Baye, J. Dohet-Eraly, P. Schoofs, Phys. Rev. A **99**, 022508 (2019)

[24] Z.D. Bai, Z.X. Zhong, Z.C. Yan, T.Y. Shi, Chin. Phys. B **30**, 023101 (2021)
[25] K. Assamagan et al., Phys. Rev. D **53**, 6065 (1996)
[26] E. Salumbides, W. Ubachs, V.I. Korobov, J. Mol. Spectrosc. **300**, 65 (2014)
[27] J. Murata, S. Tanaka, Class. Quantum Gravity **32**, 033001 (2015)
[28] F. Ficek et al., Phys. Rev. Lett. **120**, 183002 (2018)
[29] A.S. Lemos, G.C. Luna, E. Maciel, F. Dahia, Class. Quantum Gravity **36**, 245021 (2019)
[30] M. Germann et al., Phys. Rev. Research **3**, L022028 (2021)
[31] H. Banks, M. McCullough, Phys. Rev. D **103**, 075018 (2021)
[32] J. Koch, F. Scheck, Nucl. Phys. A **340**, 221 (1980)
[33] M. Trassinelli, P. Indelicato, Phys. Rev. A **76**, 012510 (2007)
[34] J.G. Fetkovich, E.G. Pewitt, Phys. Rev. Lett. **11**, 290 (1963)
[35] M.M. Block et al., Phys. Rev. Lett. **11**, 301 (1963)
[36] M.M. Block, J.B. Kopelman, C.R. Sun, Phys. Rev. **140**, B143 (1965)
[37] O.A. Zaimidoroga, R.M. Sulyaev, V.M. Tsupko-Sitnikov, Sov. Phys. JETP **25**, 63 (1967)
[38] S.N. Nakamura et al., Phys. Rev. A **45**, 6202 (1992)
[39] J.E. Russell, Phys. Rev. A **1**, 742 (1970)
[40] M. Hori et al., Phys. Rev. Lett. **94**, 063401 (2005)
[41] G.Y. Korenman, S.N. Yudin, Eur. Phys. J. D **75**, 64 (2021)
[42] J. Zatorski, K. Pachucki, Phys. Rev. A **82**, 052520 (2010)
[43] A. Sótér et al., Rev. Sci. Instrum. **85**, 023302 (2014)
[44] K. Todoroki et al., Nucl. Instr. and Meth. A **835**, 110 (2016)
[45] Y. Murakami, H. Aghai-Khozani, M. Hori, Nucl. Instrum. and Meth. A **933**, 75 (2019)
[46] H. Aghai-Khozani et al., Eur. Phys. J. Plus **127**, 125 (2012)
[47] H. Aghai-Khozani et al., Nucl. Phys. A **1009**, 122170 (2021)
[48] A. Bianconi et al., Phys. Lett. B **704**, 461 (2011)
[49] H. Aghai-Khozani et al., Nucl. Phys. A **970**, 366 (2018)
[50] M. Hori, A. Dax, Opt. Lett. **34**, 1273 (2009)
[51] H. Yamaguchi et al., Phys. Rev. A **66**, 022504 (2002)
[52] H. Yamaguchi et al., Phys. Rev. A **70**, 012501 (2004)
[53] O.I. Kartavtsev, D.E. Monakhov, S.I. Fedotov, Phys. Rev. A **61**, 062507 (2000)
[54] M. Hori et al., Phys. Rev. Lett. **87**, 093401 (2001)
[55] M. Hori et al., Phys. Rev. Lett. **91**, 123401 (2003)
[56] M. Hori et al., Phys. Rev. A **70**, 012504 (2004)
[57] B.D. Obreshkov, D.D. Bakalov, B. Lepetit, K. Szalewicz, Phys. Rev. A **69**, 042701 (2004)
[58] A. Sótér et al., to be published (2022)
[59] V.I. Korobov, Z.X. Zhong, Q.L. Tian, Phys. Rev. A **92**, 052517 (2015)
[60] M. Hori et al., Phys. Rev. Lett. **89**, 093401 (2002)
[61] J.S. Cohen, Rep. Prog. Phys. **67**, 1769 (2004)
[62] K. Tőkési, B. Juhász, J. Burgdörfer, J. Phys. B **38**, S401 (2005)
[63] X.M. Tong, K. Hino, N. Toshima, Phys. Rev. Lett. **101**, 163201 (2008)
[64] K. Sakimoto, Phys. Rev. A **91**, 042502 (2015)
[65] D. Bakalov, B. Jeziorski, T. Korona, K. Szalewicz, E. Tchoukova, Phys. Rev. Lett. **84**, 2350 (2000)
[66] S. Alighanbari, G.S. Giri, F.L. Constantin, V.I. Korobov, S. Schiller, Nature **581**, 152 (2020)

[67] S. Patra et al., Science **369**, 1238 (2020)
[68] V. Chohan et al., CERN-2014-002, Geneva, Switzerland (2014)
[69] D. Gamba et al., *ELENA COMMISSIONING*, in *Proceedings of North American Particle Accelerator Conference NAPAC2019, Lansing MI, USA* (2019), p. WEYBB1
[70] M. Hori, J. Walz, Prog. Part. Nucl. Phys. **72**, 206 (2013)
[71] M. Hori, V.I. Korobov, Phys. Rev. A **81**, 062508 (2010)
[72] A. Hirtl et al., Eur. Phys. J. A **57**, 70 (2021)
[73] V.V. Flambaum, J. Jin, D. Budker, Phys. Rev. C **103**, 054603 (2021)

Pionic hydrogen and deuterium

D. Gotta[1,*], F. D. Amaro[2], D. F. Anagnostopoulos[3], P. Bühler[4], D. S. Covita[2,5], H. Fuhrmann[4], H. Gorke[6], A. Gruber[4], M. Hennebach[1], A. Hirtl[4], T. Ishiwatari[4], P. Indelicato[7], T. S. Jensen[5,8], E.-O. Le Bigot[7], Y.-W. Liu[5], B. Manil[7], V. E. Markushin[5], J. Marton[4], M. Nekipelov[1], V. N. Pomerantsev[9], V. P. Popov[9], A. J. Rusi el Hassani[10], J. M. F. dos Santos[2], S. Schlesser[7], Ph. Schmid[4], L. M. Simons[5], Th. Strauch[1], M. Theisen[1], M. Trassinelli[11], J. F. C. A. Veloso[12], and J. Zmeskal[4]

[1]Institut für Kernphysik, Forschungszentrum Jülich, 52425 Jülich, Germany
[2]LIBPhys, Physics Department, Coimbra University, 3004-526 Coimbra, Portugal
[3]Dept. of Materials Sci. and Eng., University of Ioannina, Ioannina, 45110, Greece
[4]Stefan Meyer Institute for Subatomic Physics, Austrian Academy of Sciences, 1030 Vienna, Austria
[5]Laboratory for Particle Physics, Paul Scherrer Institute, 5232 Villigen, Switzerland
[6]Zentralinstitut für Elektronik, Forschungszentrum Jülich, 52425 Jülich, Germany
[7]LKB, Sorbonne Université, CNRS, ENS-PSL, Collège de France, 4 place Jussieu, 75005 Paris, France
[8]Institut für Theoretische Physik Universität Zürich, 8057 Zürich, Switzerland
[9]Skobeltsyn Inst. of Nuclear Physics, Lomonossov Moscow State University, 119234 Moscow, Russia
[10]Faculté des Sciences et Techniques, Université Abdelmalek Essaâdi, Tanger, Morocco
[11]Institut des NanoSciences de Paris, CNRS, Sorbonne Université, 4 place Jussieu, 75005 Paris, France
[12]I3N, Dept. of Physics, Aveiro University, 3810 Aveiro, Portugal

Abstract. The strong-interaction effects both in pionic hydrogen and deuterium atoms have been re-determined with improved precision. The hadronic shift and width in pionic hydrogen together with the hadronic shift in pionic deuterium constitute a one-fold constraint for the two independent pion-nucleon scattering lengths. Furthermore, the hadronic width in pionic deuterium measures the transition strength of s-wave pions on an isoscalar nucleon-nucleon pair which is an independent quantity not related to the pion-nucleon scattering lengths. The experiment was performed at the Paul Scherrer Institute by stopping a high-intensity low-energy pion beam in gaseous targets using the cyclotron trap. The X-rays emitted by the πH and πD atoms were analysed with a high resolution Bragg spectrometer equipped with spherically bent crystals. The pion-nucleon scattering lengths and other physical quantities extracted from the atom data are in good agreement with the results obtained from pion-nucleon and nucleon-nucleon scattering experiments and confirm that a consistent picture is achieved for the low-energy pion-nucleon sector with respect to the expectations of chiral perturbation theory.

1 Introduction

The pion-nucleon (πN) interaction at low energies plays a prominent role in the understanding of the strong interaction. Hence, $\pi N \rightarrow \pi N$ reactions and the corresponding scattering

*e-mail: d.gotta@fz-juelich.de

lengths are of fundamental interest. Therefore, a comparison of the results from precision experiments with advanced theoretical calculations is of great importance. The calculations have been performed in the frame work of effective field theories like Chiral Perturbation theory (χPT), which predict isospin-breaking effects to be of the order of a few per cent.

The experimental approach is achieved by means of high-resolution X-ray spectroscopy of pionic hydrogen and deuterium atoms. Such atomic systems are formed when pions are decelerated to kinetic energies corresponding to bound states in the Coulomb potential of a nucleus. After capture, pions undergo an atomic de-excitation cascade, where in the last steps emission of characteristic X-rays occurs. In the case of pionic hydrogen and deuterium, the strong-interaction manifests as an energy shift ϵ_{1s} and broadening Γ_{1s} of the atomic ground state $1s$ (fig. 1). Considering the energy regime of the atomic systems, the measurement of ϵ_{1s} and Γ_{1s} constitutes a scattering experiment at threshold.

The hadronic effects are related in leading order to specific πN reactions. Where $\epsilon_{1s}^{\pi H}$ is attributed to elastic scattering $\pi^- p \to \pi^- p$, $\Gamma_{1s}^{\pi H}$ is due to charge exchange $\pi^- n \to \pi^0 n$. For $\epsilon_{1s}^{\pi D}$, it is the coherent sum of $\pi^- p \to \pi^- p$ and $\pi^- n \to \pi^- n$ where, however, substantial corrections because of 3-body effects must be taken into account.

In the limit of isospin conservation, all $\pi N \to \pi N$ reactions are completely determined by only two independent real numbers, the scattering lengths corresponding to the isospin combinations $I = 1/2$ and $3/2$ of the πN system. In the isospin basis, the scattering lengths of the reaction channels read $a_{\pi^- p \to \pi^- p} = \frac{1}{3}(2a_{1/2} + a_{3/2})$, $a_{\pi^- p \to \pi^0 n} = -\frac{\sqrt{2}}{3}(a_{1/2} - a_{3/2})$, and $a_{\pi^+ p \to \pi^+ p} = a_{\pi^- n \to \pi^- n} = a_{3/2}$ leading to the so-called isospin tringle relation $a_{\pi^- p \to \pi^- p} - a_{\pi^+ p \to \pi^+ p} = -\sqrt{2} a_{\pi^- p \to \pi^0 n}$. It is convenient to use also as the two independent quantities the isoscalar and isovector scattering lengths given by $a^{\pm} \equiv (a_{\pi^- p \to \pi^- p} \pm a_{\pi^+ p \to \pi^+ p})$. The relations of the various scattering lengths to $\epsilon_{1s}^{\pi H}$, $\Gamma_{1s}^{\pi H}$, and $\epsilon_{1s}^{\pi D}$ are indicated in table 1.

In contrast hereto, $\Gamma_{1s}^{\pi D}$ is not due to scattering but to true pion absorption $\pi NN \to NN$ by an $I = 0$ nucleon-nucleon pair. Considering charge symmetry ($\pi^- d \to nn \Leftrightarrow \pi^+ d \to pp$) and detailed balance ($\pi^+ d \to pp \Leftrightarrow pp \to \pi^+ d$), except for small isospin-breaking corrections, all three reactions are described by the same transition matrix elements. Hence, the hadronic width in πD directly yields the pion-production transition strength at threshold, denoted α, when corrected for the radiative capture channels $\pi NN \to NN\gamma$ (tab. 1).

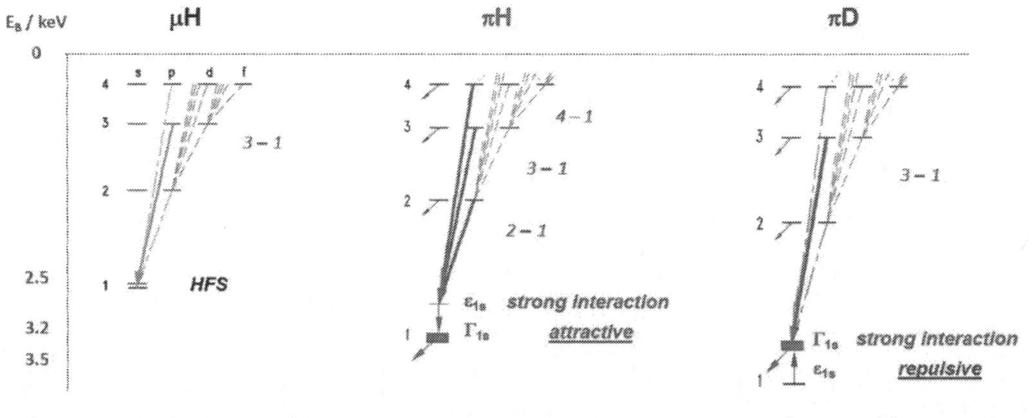

Figure 1. Scheme of low-lying atomic levels of muonic (μH) and pionic hydrogen (πH) and pionic deuterium (πD). Lyman transitions ($n - 1$) measured in this experiment are indicated. The sign convention for the hadronic shift is $\epsilon \equiv E_X - E_{QED}$, which corresponds to the change of the X-ray transition energy E_X compared to the pure electromagnetic value E_{QED}.

This paper recapitulates the final results of a long series of experiments [1] aiming at the improved determination of the strong-interaction effects both in pionic hydrogen and deuterium [2–8]. Concepts and recent theoretical efforts on low-energy πN scattering and pionic hydrogen are reviewed in [9, 10]. Properties of low Z exotic atoms, experimental methods, and the performance of the crystal spectrometer are outlined in [11, 12].

2 Experiment

In the case of muonic and pionic hydrogen, only X-rays from the K series (fig. 2 - left) are accessible within this experimental scheme (fig. 4). High resolution and high statistics measurements are confronted with various physical effects during the de-excitation cascade and need highly advanced experimental methods.

2.1 Cascade effects

The upper part of the atomic cascade is dominated by non-radiative de-excitation collision-induced reactions. Three different reactions must be considered in particular. At first, Stark mixing leads to mixing of the angular momentum states when the exotic hydrogen atom penetrates the Coulomb field of H atoms of the H_2 molecules in the target [13]. The resulting s-state admixture in the high-lying levels leads for pions to a significant reduction of X-ray line yields because of nuclear reactions (fig. 2 - right).

Secondly, additional X-ray energy shifts, thus falsifying the result for the strong-interaction shift ϵ, may occur due to molecular formation $\pi^{-}\mathrm{H}+\mathrm{H}_2 \rightarrow [(\pi^{-}pp)\cdot p]ee$ if radiative de-excitation from molecular levels exists [14, 15].

Coulomb de-excitation is a non-radiative de-excitation step $(n - n')$, where the energy is transferred into kinetic energy of the collision partners [19]. The competition of acceleration by Coulomb de-excitation and deceleration by elastic and inelastic collisions leads to a complex kinetic energy distribution at the time of X-ray emission. Advanced models like the extended standard cascade model (ESCM) follow the development of the kinetic energy distribution during de-excitation [20–23] (fig. 3 - left) which in turn required an update of the corresponding collision cross sections [24–26]. The resulting Doppler broadening can be demonstrated clearly in the line shape of the μH$(3p - 1s)$ line (fig. 3 - right). The correction

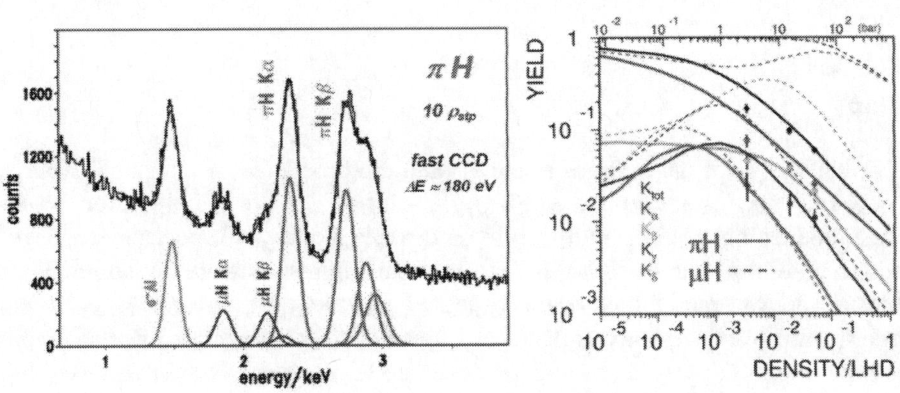

Figure 2. Left: K series of pionic hydrogen measured with a fast read-out pnCCD [16]. Muonic X-rays stem from decay muons stopped in H_2 target. Right: K yields in pionic and muonic hydrogen as calculated with the ESCM cascade model [17]. The data points are from [18].

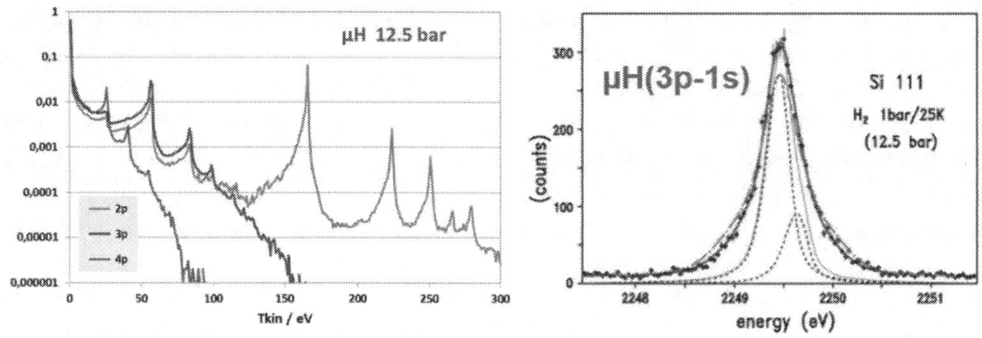

Figure 3. Left: theoretical prediction for the kinetic energy spectrum at the time of the emission of the μH$(3p-1s)$ transition. Right: measured line shape of the μH$(3p-1s)$ transition using the silicon 111 reflection. The narrow structures inside the μH line display the two hyperfine transitions as given by the spectrometer resolution only [7].

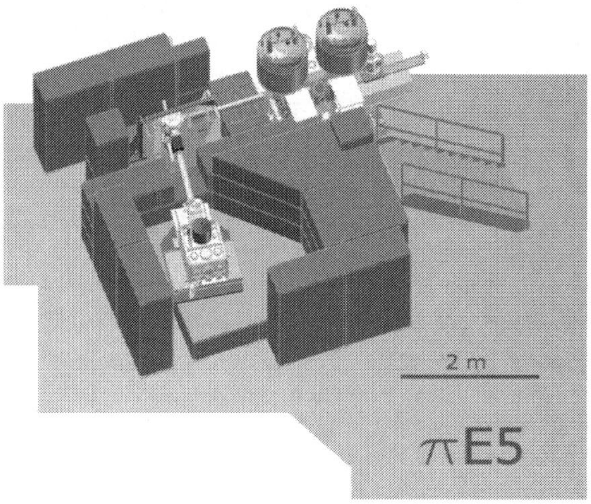

Figure 4. Set-up in the πE5 area at PSI for the πH$(4-1)$ [2] and the πD$(3-1)$ line [5]. The Bragg crystal is mounted inside a vacuum chamber (upper left) and connected to the cyclotron trap (upper right) and the CCD X-ray detector (lower left) by a vacuum system to minimize absorption losses. Bragg angles for the Si 111 reflection are $\Theta_B = 40.5°$ and $40.0°$, respectively. The roof of the concrete shielding is not shown.

for the Doppler broadening is essential for a precise determination of the strong-interaction broadening Γ.

2.2 Set-up

The energy shift ϵ_{1s} is of the order of a few eV and the broadening $\Gamma_{1s} \approx 1$ eV, whereas the $(np-1s)$ X-ray energies are around 3 keV (fig. 2 - left). The precise measurement of such a small line broadening requires high-resolution devices like crystal spectrometers, which in turn need strong X-ray sources. For this reason, precision results were not attainable before high-intensity and low-energy pion beams became available.

In order to improve the quality of the results for the strong-interaction effects compared to previous experiments [27–31], a specifically developed cyclotron trap [32] was used together with a Johann-type spectrometer equipped with spherically bent Bragg crystals as well as a large area array of charge-coupled devices (CCDs) [33]. The experiment benefited from a 4-fold higher beam intensity provided at the Paul Scherrer Institute (Villigen, Switzerland). Statistics and peak-to-background ratio were improved by a factor of about 30 and 50 by setting up a massive specially tailored concrete shielding, respectively.

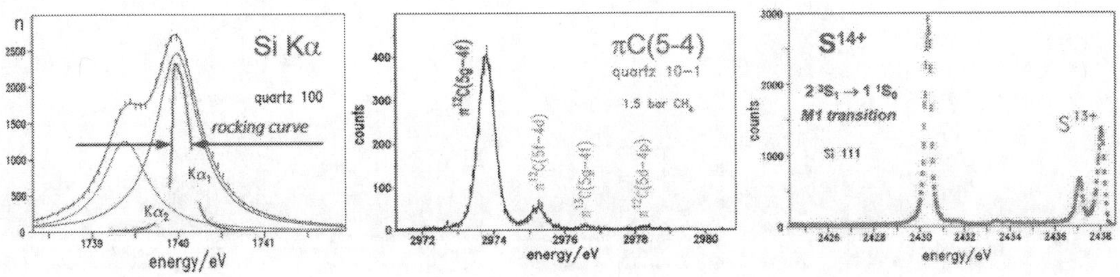

Figure 5. Left: medium Z fluorescence X-ray widths exceed the crystal resolution (rocking curve width) due to the large Auger width. Middle: pionic atoms measurements provide narrow X-ray lines but suffer from too low statistics for systematic crystal studies. Right: up to 10000 events per hour have been recorded in M1 transitions of He-like medium Z atoms [34].

Only silicon and quartz crystals meet the requirements for the resolutions needed. Thin polished slabs were spherically bent to radii of about $R_c = 3\,\text{m}$ by attaching them to glass lenses of optical quality by adhesive forces only. Resolutions of 270 – 460 meV, which are close to the theoretical limit, were achieved for the X-rays in the energy range of 2.2−3.1 keV.

The main obstacle to a precision determination of the hadronic broadening Γ_{1s} is Doppler broadening due to Coulomb de-excitation. Consequently, the ultimate knowledge of the spectrometer response function is of great importance and conventional methods to determine it are marginal (fig. 5 - left and middle).

Here, the cyclotron trap offers the unique possibility for a new calibration method when extended to operate as ECR source [35]. The ECR source yields narrow X-rays at high rates from helium-like low Z elements like sulphur, chlorine, and argon which almost coincide in energy with the pionic hydrogen and deuterium X-ray transitions (fig. 5 - right).

The kinetic energy distribution of the πH atoms develops during the atomic cascade and, in addition as a collision-induced process, must be assumed to be density dependent. As a result, the influence of the Doppler effect depends on the initial state np and the target density. Also, the molecular formation rate scales with collision probability. A density dependence of X-ray energies would demonstrate the appearance of X-ray emission from molecular states.

Therefore, three different ground-state transitions, πH($2p - 1s$), πH($3p - 1s$), and πH($4p - 1s$) (Kα, Kβ, and Kγ), have been measured at hydrogen densities between 3.9 bar and liquid (LH_2). Furthermore, the measured line shape is a convolution of spectrometer response, Doppler broadening, and the Lorentzian representing the hadronic broadening. For that reason, the measurement of the twin system μH offered the chance to study the Doppler broadening without the hadronic contribution to the line width (fig. 3 - right).

3 Analysis

3.1 Hadronic shift

The Johann-type set-up of the crystal spectrometer requires a calibration line, preferentially as close as possible in the Bragg angle of the corresponding πH line [12, 36]. For the πH($3p-1s$) transition, the energy of the πO($6h - 5g$) line almost coincides (fig. 6). This line provides a superior calibration standard compared to the fluorescence X-rays, because πO($6h - 5g$) X-rays originate from hydrogen-like pionic atoms, the energy levels of which can be calculated to $\pm$ 7 meV, where about $\pm$ 6 meV are due to the uncertainty of the pion mass. No change of the set-up is required between measurement and calibration except a gas exchange in the

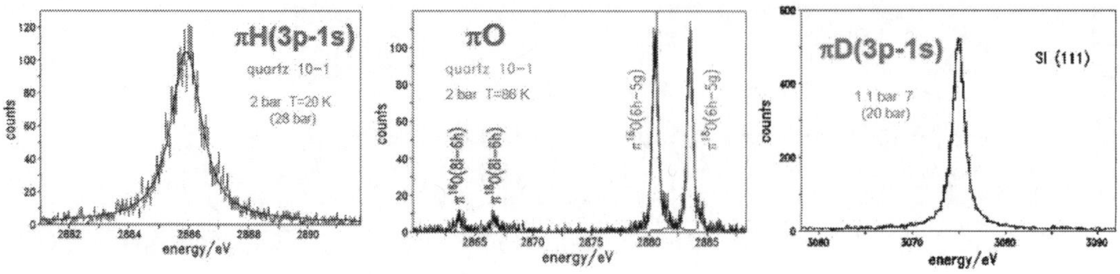

Figure 6. The energy calibration for the πH$(3p-1s)$ measurements were performed with X-rays from fully stripped pionic oxygen using the πO$(6h-5g)$ line [2]. For the πD$(3p-1s)$ line, Ga Kα fluorescence X-rays measured in 3rd order served as energy calibration [5].

target cell. At the lowest density, by using an O_2/H_2 gas mixture, both lines can be measured simultaneously.

For the πH$(2p-1s)$, πH$(4p-1s)$, and πD$(3p-1s)$ transitions measured in this experiment series [2, 5], fluorescence X-rays served as calibration standard as in previous experiments [27–31]. Results for $\epsilon_{1s}^{\pi\mathrm{H}}$ are in perfect agreement with the ones using the πO calibration. The uncertainties of the energies of fluorescence X-rays, however, amount to $\pm(27-73)$ meV for the cases considered here.

No density dependence of X-ray energies was observed for either hydrogen and deuterium, *i. e.* X-ray emission from molecular orbits does not play any role for the accuracy of the data given [2, 5].

Worthwhile to mention, that a discrepancy for $\epsilon_{1s}^{\pi\mathrm{H}}$ between the new result [2] and previous experiments disappears when readjusting the older values with a new calculation of the pure electromagnetic transition energy [37].

3.2 Hadronic broadening

The analysis of the μH, πH, and πD line shapes uses a model for the kinetic energy distribution which reduces its complexity to a few prominent components [4–8]. The model distribution consists always of a low-energy part which takes into account non accelerated or already again decelerated atoms. Depending on the initial level of the X-ray emission, up to three high-energy components were considered with the energies inspired by the fact that a few dominant Coulomb transitions dominate the Doppler broadening.

The model line profiles $LP = R \otimes L \otimes (\sum_i D_i)$ are build up by the convolution of the spectrometer response R, the natural line with L which corresponds to Γ_{1s}, and the components $(\sum_i D_i)$ modelling the kinetic energy distribution.

Both a χ^2 (frequentist) analysis and a Bayesian analysis were performed for pionic and muonic hydrogen. The non-trivial model parameters of the analyses were Γ_{1s}, the relative intensities of the Doppler components D_i, and the energies of the high-energetic components.

The use of a Bayesian approach for πH and μH was motivated by the case of pionic deuterium. There, the occurrence of a bias in a frequentist method [38], influencing the extracted values for $\Gamma_{1s}^{\pi\mathrm{D}}$, has been studied in detail [5]. Such a bias stems from the — principally unknown — difference of the probability distribution of the data itself and the one assumed for the model. An estimate for the bias of the parameter $\Gamma_{1s}^{\pi\mathrm{D}}$ was obtained by performing with the model as used in the analysis a series of Monte-Carlo simulations to generate the X-ray line shape and, after analysis, comparing input and output parameters. This lengthy procedure may be circumvented by using Bayesian methods, which are supposed to be free from bias effects in parameter estimation [39]. Furthermore, comparing the results of the two analysis methods establishes an independent consistency check.

Both the frequentist as well as the Bayesian analyses clearly show that a distinctive low-energy component of the kinetic energy distribution exists containing about 2/3 of the total intensity. As expected, it is impossible to resolve the complex structure of the kinetic energy spectrum at higher energies in the presence of a hadronic broadening with the data given. However, it can be effectively modelled by one or two high-energy components and the details of these components are not critical for the numerical result of $\Gamma_{1s}^{\pi H}$ but limits the accuracy to about 3%. To summarize the current situation one can assert that the theoretical prediction for the kinetic energy distributions and the experimental results do not show any drastic inconsistencies. The weighted average for the strong-interaction effects over the various transitions and target densities measured is given in table 1.

It is worth mentioning that in pionic deuterium no Doppler broadening due to Coulomb de-excitation could be identified within the experimental uncertainties [5]. A theoretical explanation for such behaviour has been provided recently by cascade theory [40].

Table 1. Results for the strong-interaction effects ϵ_{1s} and Γ_{1s} in pionic hydrogen and deuterium (in meV) and their relation to πN scattering and absorption reactions as well as to the isoscalar and isovector scattering length a^+ and a^- and the threshold pion production strength α.

$\epsilon_{1s}^{\pi H}$	$\Gamma_{1s}^{\pi H}$	$\epsilon_{1s}^{\pi D}$	$\Gamma_{1s}^{\pi D}$
7085.8 ± 9.6 [2]	856 ± 27 [8]	-2356 ± 31 [5]	1171^{+23}_{-49} [5]
$\pi^- p \to \pi^- p$	$\pi^- p \to \pi^0 n$	$\pi^- p \to \pi^- p + \pi^- n$	$\Gamma_{\pi^- d \to nn + nn\gamma}$
$\propto (a^+ + a^-) + ...$	$\propto (a^-)^2 + ...$	$\propto 2a^+ + ...$	α

3.3 Results

The three independent experimental results for the strong-interaction shifts in πH and πD together with the hadronic broadening in πH yield a onefold constraint for any combination of the two independent πN scattering length. As shown in fig. 7 - left, a consistent picture is achieved by applying the corrections as worked out in the framework of χPT [10, 41, 42]. The uncertainty for the scattering lengths as determined from the two shifts is mainly attributed to the poor knowledge of LECs, whereas in the case of the hadronic broadening the experimental error is about twice the one of the chiral corrections.

In addition to the scattering length of the reactions $\pi^- p \to \pi^- p$ and $\pi^- n \to \pi^0 n$, the scattering length for the reaction $\pi^+ p \to \pi^+ p$ is accessible via the triangle relation applying appropriate chiral corrections. Furthermore, the πN coupling constant $g_c^2/4\pi$ [43], the pion photo-production strength at threshold $E_{0^+}^{thr}$ [44], or the πN σ-term $\sigma_{\pi N}$ [45] also are related to the isoscalar and isovector scattering lengths. A comparison of the these quantities derived from pionic-atom data and other sources is shown in fig. 8. In general, data from various sources are highly consistent when analysed within the framework of χPT [42, 46, 47]. For $\sigma_{\pi N}$, a discrepancy of about 20 MeV remains between the experiment based values [8, 47–49] and lattice calculations [50–53] which, however, may be removed by including $N\pi$ and $N\pi\pi$ contributions [54].

The result for the pion-production strength α, extracted from the strong-interaction width in πD, is compared to the results from pion-production experiments in fig. 7 - right. The theoretical approach within the frame work of χPT suffers again from the scarce precision of some LECs [57].

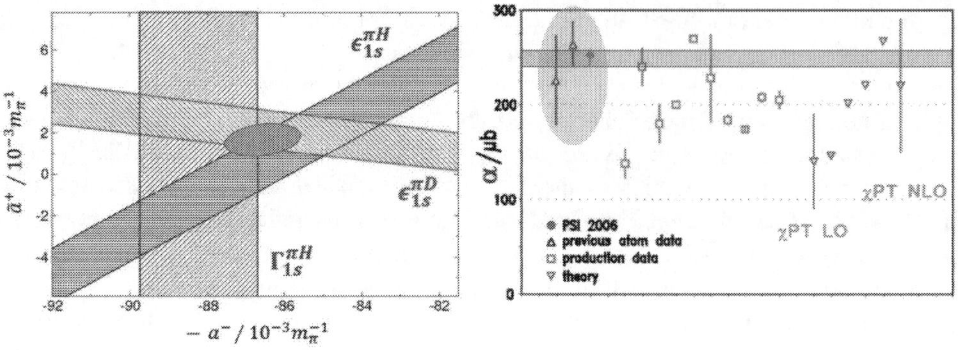

Figure 7. Left: constraints (bands) and combined result (ellipse) for the isoscalar and isovector πN scattering lengths $\tilde{a}^+$ and a^- as derived from $\epsilon_{1s}^{\pi H}$, $\epsilon_{1s}^{\pi D}$, and $\Gamma_{1s}^{\pi H}$ [8]. Right: comparison of results for pion-production strength α at threshold on isoscalar NN pairs. The horizontal band represents the precision of the most recent result for $\Gamma_{1s}^{\pi D}$ [5].

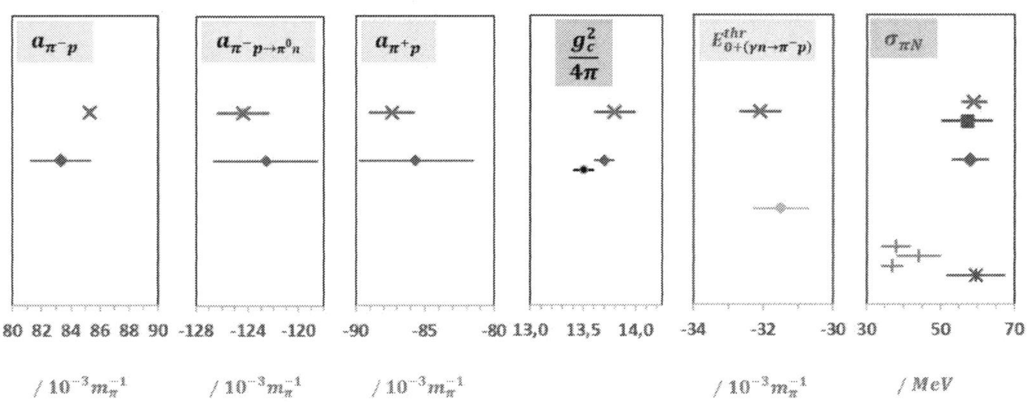

Figure 8. Comparison of parameters derived from pionic hydrogen and deuterium strong-interaction effects $\epsilon_{1s}^{\pi H}$, $\epsilon_{1s}^{\pi D}$, and $\Gamma_{1s}^{\pi H}$ (× [2, 3, 5, 8]) and results from other sources (■ higher Z pionic-atoms in medium fit [49], ♦ πN scattering analysis based on Roy-Steiner equations [48], • NN scattering [55], ▲ Photo production [56], + lattice calculations [50–53], ✱ lattice calculations including low-lying $N\pi$ and $N\pi\pi$ states [54]).

3.4 Conclusion and outlook

In summary, recent πH, πD, and low-energy πN scattering data are quantitatively consistent when analysed within the framework of χPT. In view of the difficulties arising from the line-shape corrections due to Coulomb de-excitation, future experiments should preferably use X-ray transitions from higher np initial states where the Doppler broadening is expected to be smaller. Since the accelerator currents improve continuously with time, measurements with sufficient statistics for the $(4-1)$ or even $(5-1)$ transition will become feasible. In this respect it is important to emphasize that the coverage of a sufficiently large energy interval around the transition is essential to fix the background level on both sides of the line. In this way, an accuracy of 1% or better can be achieved for the hadronic width of pionic hydrogen.

References

[1] PSI proposals R-98.01 and R-06.03, http://collaborations.fz-juelich.de/ikp/exotic-atoms
[2] M. Hennebach et al., Eur. Phys. J. A **50**, 190 (2014)
[3] M. Hennebach et al., Eur. Phys. J. A **55**, 24(E) (2019)
[4] T. Strauch et al., Phys. Rev. Lett. **104**, 142503 (2010)
[5] T. Strauch et al., Eur. Phys. J. A **47**, 88 (2011)
[6] D.S. Covita et al., Phys. Rev. Lett. **102**, 023401 (2009)
[7] D.S. Covita et al., Eur. Phys. J. D **72**, 72 (2018)
[8] A. Hirtl et al., Eur. Phys. J. A **57**, 70 (2021)
[9] J. Gasser, V.E. Lyubovitzkij, A. Rusetski, Phys. Rep. **456**, 167 (2008)
[10] M. Hoferichter et al., Phys. Rep. **625**, 1 (2016)
[11] D. Gotta, Prog. Part. Nucl. Phys **52**, 133 (2004)
[12] D.E. Gotta, L.M. Simons, Spectrochim. Acta B **120**, 9 (2016)
[13] M. Leon, H. Bethe, Phys. Rev. **127**, 636 (1962)
[14] S. Jonsell, J. Wallenius, P. Froelich, Phys. Rev. A **59**, 3440 (1999)
[15] S. Kilic, J.P. Karr, L. Hilico, Phys. Rev. A **70**, 042506 (2004)
[16] H. Gorke et al., AIP Conf. Proc. **793**, 341 (2005)
[17] T.S. Jensen, V.E. Markushin, Hyperfine Int. **138**, 71 (2001)
[18] A.J. Rusi el Hassani et al., Z. Phys. A **351**, 113 (1995)
[19] L. Bracchi, G. Fiorentini, Nuovo Cim. A **43**, 9 (1978)
[20] V.E. Markushin, Phys. Rev A **50**, 1137 (1994)
[21] T.S. Jensen, V.E. Markushin, Lec. Notes Phys. **627**, 37 (2003)
[22] T.S. Jensen, V.P. Popov, V.N. Pomerantsev, arXiv:0712.3010 [nucl-th] (2007)
[23] V.P. Popov, V.N. Pomerantsev, Hyperfine Int. **209**, 75 (2012)
[24] T.S. Jensen, V.E. Markushin, Eur. Phys. J. D **21**, 165 (2002)
[25] V.N. Pomerantsev, V.P. Popov, Phys. Rev. A **73**, 040501(R) (2006)
[26] V.P. Popov, V.N. Pomerantsev, Phys. Rev. A **86**, 052520 (2012)
[27] D. Sigg et al., Phys. Rev. Lett. **75**, 3245 (1995)
[28] D. Sigg et al., Nucl. Phys. A **609**, 269 (1996)
[29] D. Sigg et al., Nucl. Phys. A **617**, 526 (1997)
[30] H.C. Schröder et al., Phys. Lett. B **469**, 25 (1999)
[31] H.C. Schröder et al., Eur. Phys. J. C **21**, 473 (2001)
[32] D. Gotta, L.M. Simons, SciPost Phys. Proc. **5**, 013 (2021)
[33] N. Nelms et al., Nucl. Instr. Meth. A **484**, 419 (2002)
[34] D.F. Anagnostopoulos et al., Nucl. Instr. and Meth. A **545**, 217 (2005)
[35] S. Biri, L. Simons, D. Hitz, Rev. Sci. Instrum. **71**, 1116 (2000)
[36] M. Trassinelli et al., Phys. Lett. B **759**, 583 (2016)
[37] S. Schlesser et al., Phys. Rev. C **84**, 015211 (2011)
[38] U.C. Bergmann, K. Riisager, Nucl. Instr. Methods A **489**, 444 (2002)
[39] R.D. Cousins, Am. J. Phys. **63**, 398 (1995)
[40] V.P. Popov, V.N. Pomerantsev, Phys. Rev. A **95**, 022506 (2017)
[41] V. Baru et al., Nucl. Phys. A **872**, 69 (2011)
[42] V. Baru et al., Phys. Lett. B **694**, 473 (2011)
[43] M.L. Goldberger, H. Miyazawa, R. Oehme, Phys. Rev. **99**, 986 (1955)
[44] N.M. Kroll, M.A. Ruderman, Phys. Rev. **93**, 233 (1954)

[45] J. Gasser, H. Leutwyler, M.E. Sainio, Phys. Lett. B **253**, 252 (1991)
[46] V. Bernard, N. Kaiser, U.G. Meißner, Phys. Lett. B **383**, 116 (1996)
[47] M. Hoferichter, J. Ruiz de Elvira, B. Kubis, U.G. Meißner, Phys. Rev. Lett. **115**, 092301 (2015)
[48] J. Ruiz de Elvira et al., J. Phys. G **45**, 024001 (2018)
[49] E. Friedman, A. Gal, Phys. Lett. B **792**, 340 (2019)
[50] S. Dürr et al., Phys. Rev. Lett. **116**, 172001 (2016)
[51] S. Dürr et al., Phys. Rev. Lett. **116**, 172001 (2016)
[52] A. Abdel-Rehim et al., Phys. Rev. Lett. **116**, 252001 (2016)
[53] S. Borsanyi et al., arXiv:2007.03319v1 [hep-lat] (2020)
[54] R. Gupta et al., Phys. Rev. Lett. **127**, 242002 (2021)
[55] M.C.M. deSwart, J. J. andRentmeester, R.G.E. Timmermans, PiN Newsl. **13**, 96 (1997)
[56] M.A. Kovash, PiN Newsl. **12**, 51 (1997)
[57] V. Lensky et al., Eur. Phys. J. A **27**, 37 (2006)

Low energy kaon-nuclei interaction studies at DAΦNE

Kristian Piscicchia[1,2,*], *Michael* Cargnelli[3], *Raffaele* Del Grande[4,2], *Laura* Fabbietti[5,4], *Johann* Marton[3], *Pawel* Moskal[6], *Alessandro* Scordo[2], *Diana* Sirghi[2], *Magdalena* Skurzok[2,6], *Oton* Vazquez Doce[2], *Slavomir* Wycech[7], *Johann* Zmeskal[3], *Paolo* Branchini[8], *Eryk* Czerwinski[6], *Veronica* De Leo[9,10], *Erika* De Lucia[2], *Alessandro* Di Cicco[11,8], *Eleonora* Diociaiuti[2], *Paolo* Fermani[2], *Salvatore* Fiore[12,10], *Matteo* Martini[2,13], *Elena* Perez Del Rio[9,10], *Andrea* Selce[11,8], *Michal* Silarski[6], and *Catalina* Curceanu[2]

[1]Centro Ricerche Enrico Fermi - Museo Storico della Fisica e Centro Studi e Ricerche "Enrico Fermi", 00184 Rome, Italy
[2]INFN - Laboratori Nazionali di Frascati, 00044 Frascati, Italy
[3]Stefan-Meyer-Institut für Subatomare Physik, 1030 Wien, Austria
[4]Physik Department E12, Technische Universität München, 85748 Garching, Germany
[5]Excellence Cluster "Origin and Structure of the Universe", 85748 Garching, Germany
[6]Institute of Physics, Jagiellonian University, 30-348 Cracow, Poland
[7]National Centre for Nuclear Research, 02-093 Warsaw, Poland
[8]INFN Sezione di Roma Tre, Roma, Italy
[9]Dipartimento di Fisica dell'Università "Sapienza", Roma, Italy
[10]INFN Sezione di Roma, Roma, Italy
[11]Dipartimento di Matematica e Fisica dell'Università "Roma Tre", Roma, Italy
[12]ENEA, Department of Fusion and Technology for Nuclear Safety and Security, Frascati (RM), Italy
[13]Dipartimento di Scienze e Tecnologie applicate, Università "Guglielmo Marconi", Roma, Italy

Abstract. The aim of the AMADEUS experiment is to investigate the low-energy antikaon interaction with nucleons and nuclei, exploiting the unique low-momentum beam of kaons produced by the DAΦNE collider at LNF-INFN, to constrain hadronic nuclear physics models in the strangeness -1 sector.
As a first step the data collected in 2004/2005 by the KLOE collaboration, consisting in a complex of K^- absorptions in H, ^{4}He, ^{9}Be and ^{12}C, was analyzed, leading to the first invariant mass spectroscopic study with very low momentum (about 100 MeV) in-flight K^- captures. A dedicated pure Carbon target was also implemented in the central region of the KLOE detector, providing a high statistic reference sample of pure at-rest K^- nuclear interaction. The first measurement of the non-resonant transition amplitude $|T_{K^-n\to\Lambda\pi^-}|$ at $\sqrt{s}$ =33 MeV below the $\overline{K}N$ threshold is presented, in relation with the $\Lambda(1405)$ properties studies.

1 INTRODUCTION

A specimen of the analyses which are being performed by the AMADEUS collaboration will be described. We will be in particular concerned on the investigation of the resonant and non-resonant transition amplitudes, below the $\overline{K}N$ threshold. These studies take advantage of

*e-mail: kristian.piscicchia@cref.it

the unique low momentum kaons beam produced by the DAΦNE collider [1] at the Frascati National Laboratory of INFN. At-rest and low-mometum in-flight ($p_K \sim 127$ MeV/c) K^- mesons captures on ^{4}He and ^{12}C nuclear targets are exploited. The study is aimed to provide constraints to the non-perturbative strong interaction models in the strangeness sector.

Quantitative information on the yield and the spectral shape of the non-resonant antikaon-nucleon absorption transition amplitude, below the $\overline{K}N$ threshold, giving rise to a hyperon-pion final state, is a useful reference to extract the Λ(1405) properties and to disentangle this state from the competing non-resonant $(\Sigma\pi)^0$ production in $\overline{K}N$ absorption experiments.

The antikaon-nucleon strong interaction is dominated by the emergence of the (1405) (isospin I=0) state below the $\overline{K}N$ threshold. Phenomenological potential models (see e.g. Refs. [2, 3]) interpret the (1405) as a pure $\overline{K}N$ bound state with a Binding Energy of about 27 MeV. According to chiral unitary models [4–10] the (1405) emerges from the interference of two poles, as a molecular meson-baryon state. A lower pole, with a mass of about 1390 MeV, is mainly coupled to the $\Sigma\pi$ decay channel, a higher mass pole, which is mainly coupled to the $\overline{K}N$ channel, is located around 1420 MeV (see also [11]). The shape of the measured $(\Sigma\pi)^0$ invariant mass distribution should thus depend on both the decay channel (as a consequence of the isospin interference term, which contributes with opposite sign to the $\Sigma^\pm\pi^\mp$ cross sections and vanishes for $\Sigma^0\pi^0$) and on the production channel. In this scenario $\overline{K}N$ absorption represents the golden channel for investigating the predicted high mass pole of the Λ(1405). The experimental characterisation of the antikaon-nucleon amplitude below threshold is a good constraint for the chiral predictions (see e.g. [6, 12]) since the real and imaginary parts of the calculated transition amplitudes are strongly model dependent in the sub-threshold region both in I=0 and I=1.

The analysed data sample collected by the KLOE collaboration in 2004/2005 [13] permits investigating both at rest ($p_K \sim 0$) and in-flight low momentum K^- nuclear absorptions. In the case of interactions at-rest the underlying mechanism consists in the capture of the strange meson in a highly excited atomic state and a successive cascade to low-lying states, followed by the K^- nuclear capture. Besides the interactions at-rest, an important contribution from in-flight K^- nuclear captures was already characterised by the AMADEUS collaboration in previous works (see e.g. [14]). In this case the kaon crosses the electron cloud, and is directly absorbed by the nucleus. We will show that low-momentum kaon induced reactions in-flight can give important information on the Λ(1405) high mass pole properties.

2 DATA SAMPLES

We are presently analyzing two complementary samples of data. The first one is represented by the data collected by the KLOE collaboration during the 2004/2005 data taking campaign, and corresponds to an integrated luminosity of about 1.74 fb^{-1}. In this case the KLOE detector is used as an active target, the hadronic interaction of negative kaons with the materials of the apparatus being investigated; in particular K^-^{9}Be absorptions in the DAΦNE Beryllium thin cylindrical layer and the DAΦNE aluminated beryllium pipe, K^-^{12}C and K^-H absorptions in the KLOE Drift Chamber [15] (DC) inner wall (aluminated carbon fiber), K^-^{4}He in the DC gas, filled with a mixture of Helium and Isobutane (in volume: 90% ^{4}He and 10% C_4H_{10}). This sample provides extremely rich experimental information, since it contains K^- hadronic captures both at-rest and in-flight.

A second dedicated measurement was performed in 2012, with a high purity carbon (graphite) target, which was installed between the beam pipe and the KLOE DC inner wall. Aim of the measurement was to collect a reference sample of pure K^- absorption at-rest in ^{12}C. The geometry of the target was optimized to maximize the kaon stopping power (technical details can be found in Ref. [14]). The total reconstructed integrated luminosity amounts to about 37 pb^{-1}.

The detailed event selection and particle identification procedure, for the analyses which is be described in this work, is reported in Ref. [14].

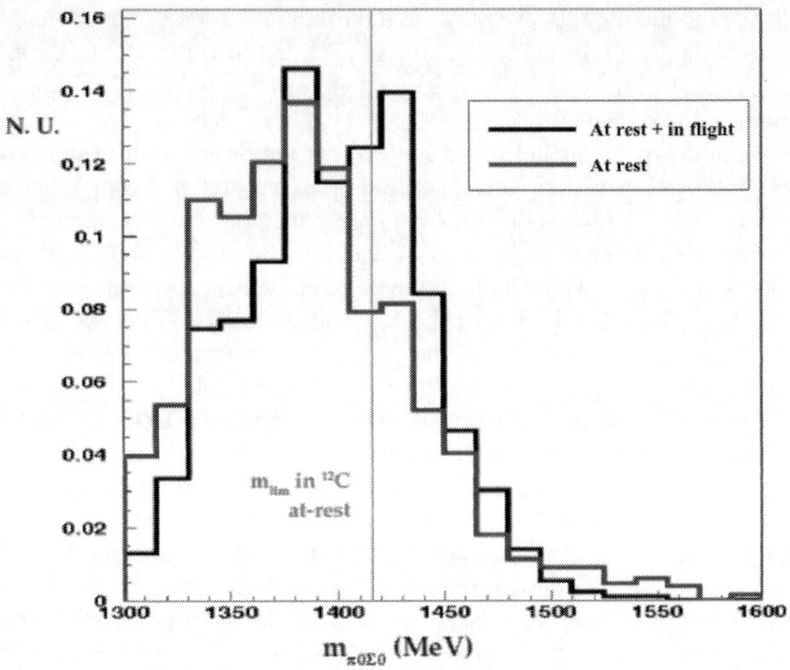

Figure 1. *(Colour online.)* $m_{\Sigma^0\pi^0}$ *invariant mass distribution from* K^- *captures in the KLOE DC wall (black curve) and pure carbon graphite target (blue curve).*

3 RESONANT AND NON RESONANT Yπ TRANSITION AMPLITUDES BELOW THE $\overline{K}$N THRESHOLD

The investigation of the Λ(1405) properties in K^- induced reactions on light nuclear targets is strongly biased by the kinematic threshold, determined by the absorbing nucleon binding energy. For K^- captures at-rest on ^{4}He and ^{12}C the $m_{\Sigma\pi}$ invariant mass thresholds are about 1412 MeV and 1416 MeV respectively. The $m_{\Sigma\pi}$ range lying immediately below the $\overline{K}$N threshold can be accessed, by exploiting the in-flight K^- capture process, taking advantage of the kaon kinetic energy contribution. For kaon momenta of 100 MeV/c the $m_{\Sigma\pi}$ threshold is shifted upwards by about 10 MeV, thus opening the mass window where the high mass pole of the Λ(1405) is predicted.

In Fig. 1 (black distribution) the $\Sigma^0\pi^0$ invariant mass spectrum from K^- captures in the KLOE DC wall is shown [16]. $\Sigma^0\pi^0$ represents the best signature of the Λ(1405) resonance, since it does not contain isospin $I = 1$ background. The $m_{\Sigma^0\pi^0}$ spectrum is compared with the corresponding distribution of K^- captures at-rest in the pure Carbon target (fig. 1 blue distribution). The blue and the black distributions are normalised to unity. In fig. 1 the red line indicates the energy threshold corresponding to K^- absorption in ^{12}C at-rest. A rich sample of in-flight K^- ^{12}C captures can be identified above the red line. The next steps of the data analysis, presently ongoing, consist in the subtraction of the non-resonant $\Sigma^0\pi^0$ formation contribution, which is due to the K^- absorption on hydrogen, contained in the carbon fibre, and in a comparison of the obtained spectrum with models of $\Lambda(1405)$ production in K^- captures in-flight on ^{12}C.

The real and imaginary parts of the non-resonant coupled channels $K^-n\rightarrow \Lambda\pi/\Sigma\pi$ scattering amplitudes, were calculated on the basis of several chiral SU(3) meson-baryon coupled channels interaction models (we will use the following notation: Barcelona (BCN) [6], Kyoto-Munich (KM) [7], Prague (P) [8], Murcia (M1 and M2) [9], Bonn (B2 and B4) [10], see also Refs. [17, 18]). The amplitudes values strongly differ in the $\overline{K}N$ sub-threshold region depending on the adopted model.

In order to pin down the model to be adopted for the description of the observed $\Sigma^0\pi^0$ spectra, the $K^-n\rightarrow \Lambda\pi^-$ transition was measured by exploiting K^-n single nucleon absorptions in ^{4}He. Since the $\Sigma^-(1385)$ (isospin $I = 1$) resonance is well known, the corresponding non-resonant transition amplitude ($|T_{K^-n\rightarrow\Lambda\pi^-}|$) can be extracted, and used to test the chiral predictions below threshold. The details of the particle identification and event selection are given in Ref. [12]. The contribution of the Σ particle conversion events, to the measured sample, was minimised by selecting, in the p_{π^-} vs p_Λ scatterplot, the phase space region which is populated by the correlated direct $\Lambda\pi^-$ production. The phase space cut was optimised based on MC simulations (performed with the standard KLOE GEANT digitization (GEANFI [19])), according to the calculations reported in [20].

In order to extract the ratio of the resonant over non-resonant $\Lambda\pi^-$ production, the measured $p_{\Lambda\pi}$, $m_{\Lambda\pi}$ and $\cos(\theta_{\Lambda\pi})$ (total hyperon-pion momentum, invariant mass and the angle between the Λ and π momenta) distributions were fitted. The main background is represented by the contribution of K^- ^{12}C absorptions as a consequence of the presence of Isobutane molecules in the DC gas. In order to account for this contribution an experimental sample of K^- ^{12}C absorptions is used in the global fit. The carbon sample is obtained by selecting kaons interactions in the DC entrance wall, adopting the same criteria as for the gas sample. The modulus of the amplitude of the non-resonant processes, the ratio of the resonant to non-resonant processes, the modulus of the amplitude of the conversions and the contribution of the K^- captures on Carbon are considered as free parameters in the fit.

The results of the fit are summarised in Table I. (In Ref [12] the corresponding fit is shown in Fig. 3). The systematic errors are estimated by varying independently the selection cuts such as to increase or decrease the $\Lambda\pi^-$ statistics by 15% with respect to the optimized selection. The chi-square of the fit is $\chi^2/ndf = 151/148$. The ratio ($\frac{\text{RES-if}}{\text{NR-if}}$) for the $\Lambda\pi^-$ production in-flight is found to be smaller than the corresponding ratio at-rest. This is not surprising as the K^-n interaction in-flight occurs about $\sqrt{s}$ =20 MeV below the $\overline{K}N$ threshold; the corresponding reaction at-rest occurs about $\sqrt{s}$ =33 MeV below the threshold (see Ref. [20]), nearer to the resonance which lays about 49 MeV below the $\overline{K}N$ threshold. The systematic uncertainty on the resonant to non-resonant ratio for the in-flight reactions, prevents from extracting the modulus of the non-resonant transition amplitude in-flight. The corresponding transition amplitude modulus at-rest, extracted by comparing Eq. (14) and Eq. (20) in Ref. [20] with the experimental result ($\frac{\text{RES-ar}}{\text{NR-ar}}$) is:

$$|T_{K^-n\rightarrow\Lambda\pi^-}| = (0.334 \pm 0.018\,\text{stat}^{+0.034}_{-0.058}\text{syst})\,\text{fm}\,. \quad (1)$$

4 DISCUSSION

In this work a summary of the measurement of the isospin $I = 1$ non-resonant transition amplitude $|T_{K^-n\rightarrow\Lambda\pi^-}|$ at $\sqrt{s}$ =33 MeV below the $\overline{K}N$ threshold is presented. The result of this analysis (with combined statistical and systematic errors) is shown in Fig. 2 and compared

Channels	**Ratio/yield**	σ_{stat}	σ_{syst}
RES-ar/NR-ar	0.39	± 0.04	$^{+0.18}_{-0.07}$
RES-if/NR-if	0.23	± 0.03	$^{+0.23}_{-0.22}$
NR-ar	12.0 %	± 1.7 %	$^{+2.0}_{-2.8}$ %
NR-if	19.2 %	± 4.4 %	$^{+5.9}_{-3.3}$ %
$\Sigma \to \Lambda$ conv.	2.2 %	± 0.3 %	$^{+1.6}_{-0.8}$ %
K^{-12}C capture	57.0 %	± 1.2 %	$^{+2.2}_{-3.2}$ %

Table 1. Resonant to non-resonant ratios and amplitudes of the various channels extracted from the fit of the $\Lambda\pi^-$ sample. We indicate with RES and NR resonant and non-resonant processes respectively, the abbreviations ar and if are used for captures at-rest and in-flight. The statistical and systematic errors are also shown. See text for details.

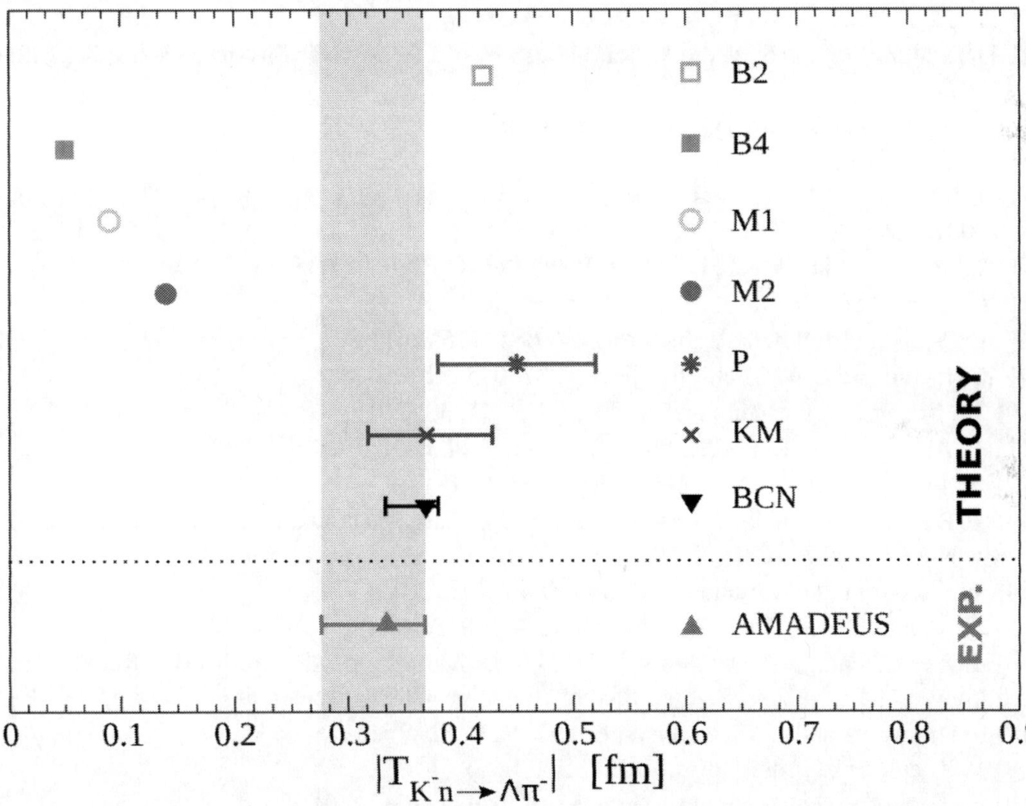

Figure 2. *(Color online) Modulus of the non-resonant transition amplitude for the $K^-n \to \pi^-$ process obtained by AMADEUS in Ref. [20], compared with theoretical predictions: Barcelona (BCN) [6], Kyoto-Munich (KM) [7], Prague (P) [8], Murcia (M1 and M2) [9], Bonn (B2 and B4) [10].*

with the theoretical predictions (BCN [6], KM [7], P [8], M1 and M2 [9], B2 and B4 [10]), re-scaled for the $K^-n \to \Sigma\pi$ transition probabilities. The presented result allows an extrapolation to the un-physical region and can be used to test models of S-wave interaction; moreover such extrapolation can be used to constrain the corresponding $(\Sigma\pi)^0$ non-resonant background for the extraction of the $\Lambda(1405)$ properties in $\overline{K}N$ absorption experiments. In this direction an analogous, theoretical and experimental, analysis of the in medium K^-p absorption is ongoing.

ACKNOWLEDGMENTS

We acknowledge the KLOE/KLOE-2 Collaboration for their support and for having provided us with the data and the tools to perform the analysis presented in this paper. Part of this work was supported by Minstero degli Affari Esteri e della Cooperazione Internazionale, Direzione Generale per la Promozione del Sistema Paese (MAECI), Strange Matter project PRG00892; EU STRONG-2020 project (grant agreement No 824093); Polish National Science Center through grant No. UMO-2016/21/D/ST2/01155.

References

[1] A. Gallo *et al.*, Conf. Proc. C **060626**, 604 (2006)
[2] L. R. Staronski and S. Wycech, Journal of Physics G: Nuclear Physics, **13**, no. 11, 1361 (1987).
[3] Y. Akaishi and T. Yamazaki, Phys. Rev. C **65**, 044005 (2002).
[4] J. A. Oller and U. G. Meissner, Phys. Lett. B **500**, 263-272 (2001).
[5] D. Jido, J. A. Oller, E. Oset, A. Ramos and U. G. Meissner, Nucl. Phys. A **725**, 181-200 (2003).
[6] A. Feijoo, V. Magas and A. Ramos, Phys. Rev. C **99**, no.3, 035211 (2019).
[7] Y. Ikeda, T. Hyodo and W. Weise, Nucl. Phys. A **881**, 98 (2012).
[8] A. Cieplý and J. Smejkal, Nucl. Phys. A **881**, 115 (2012).
[9] Z. H. Guo and J. A. Oller, Phys. Rev. C **87**, 035202 (2013).
[10] M. Mai and U. G. Meißner, Eur. Phys. J. A **51**, 30 (2015).
[11] S. Acharya *et al.* [ALICE], Phys. Rev. Lett. **124**, no.9, 092301 (2020).
[12] K. Piscicchia *et al.*, Phys. Lett. B **782**, 339 (2018)
[13] F. Bossi, E. De Lucia, J. Lee-Franzini, S. Miscetti, M. Palutan [KLOE Collaboration], Riv. Nuovo Cim. **31**, 531 (2008)
[14] C. Curceanu *et al.*, Acta Phys. Polon. B **46**, 203 (2015)
[15] M. Adinolfi *et al.*, Nucl. Instrum. Meth. A **488**, 51 (2002)
[16] K. Piscicchia, PhD thesis (2013), Università degli Studi di Roma Tre, http://www.infn.it/thesis/PDF/getfile.php?filename=7097-Piscicchia-dottorato.pdf
[17] J. Hrtankova and J. Mares, Phys. Rev. C **96**, 015205 (2017)
[18] A. Cieplý *et al.*, Nucl. Phys. A **954**, 17 (2016)
[19] F. Ambrosino *et al.*, Nucl. Instrum. Meth. A **534**, 403 (2004)
[20] K. Piscicchia, S. Wycech and C. Curceanu, Nucl. Phys. A **954**, 75 (2016)

Minimizing plasma temperature for antimatter mixing experiments

E. D. Hunter[1*], C. Amsler[1], H. Breuker[2], S. Chesnevskaya[1], G. Costantini[3], R. Ferragut[4,5], M. Giammarchi[5], A. Gligorova[1], G. Gosta[3], H. Higaki[6], Y. Kanai[7], C. Killian[1], V. Kletzl[1], V. Kraxberger[1], N. Kuroda[8], A. Lanz[1], M. Leali[3], V. Mäckel[1], G. Maero[5,9], C. Malbrunot[10], V. Mascagna[3], Y. Matsuda[8], S. Migliorati[3], D. J. Murtagh[1], Y. Nagata[11], A. Nanda[1], L. Nowak[10], E. Pasino[5,9], M. Romé[5,9], M. C. Simon[1], M. Tajima[7], V. Toso[4,5], S. Ulmer[2], U. Uggerhøj[12], L. Venturelli[3], A. Weiser[1], E. Widmann[1], T. Wolz[10], Y. Yamazaki[2], and J. Zmeskal[1]
(The ASACUSA-Cusp Collaboration)

[1]Stefan Meyer Institute
[2]Ulmer Fundamental Symmetries Laboratory, RIKEN
[3]Dipartimento di Ingegneria dell'Informazione, Università degli Studi di Brescia and INFN Pavia
[4]Politechnico di Milano
[5]INFN Milano
[6]Graduate School of Advanced Science and Engineering, Hiroshima University
[7]Nishina Center for Accelerator-Based Science, RIKEN
[8]Institute of Physics, the University of Tokyo
[9]Dipartimento di Fisica, Università degli Studi di Milano
[10]Experimental Physics Department, CERN
[11]Department of Physics, Tokyo University of Science
[12]Department of Physics and Astronomy, Aarhus University

Abstract. The ASACUSA collaboration produces a beam of antihydrogen atoms by mixing pure positron and antiproton plasmas in a strong magnetic field with a double cusp geometry. The positrons cool via cyclotron radiation inside the cryogenic trap. Low positron temperature is essential for increasing the fraction of antihydrogen atoms which reach the ground state prior to exiting the trap. Many experimental groups observe that such plasmas reach equilibrium at a temperature well above the temperature of the surrounding electrodes. This problem is typically attributed to electronic noise and plasma expansion, which heat the plasma. The present work reports anomalous heating far beyond what can be attributed to those two sources. The heating seems to be a result of the axially open trap geometry, which couples the plasma to the external (300 K) environment via microwave radiation.

1 Introduction

The ASACUSA-Cusp project aims to measure the ground state hyperfine splitting of antihydrogen in a magnetic field-free region with a precision of 1 ppm [1]. The hyperfine measure-

*e-mail: eric.david.hunter@cern.ch

ment is favored among possible probes of CPT violation because of the way it depends on the spin of the positron-antiproton system [2]. The experimental concept is sketched in Fig. 1.

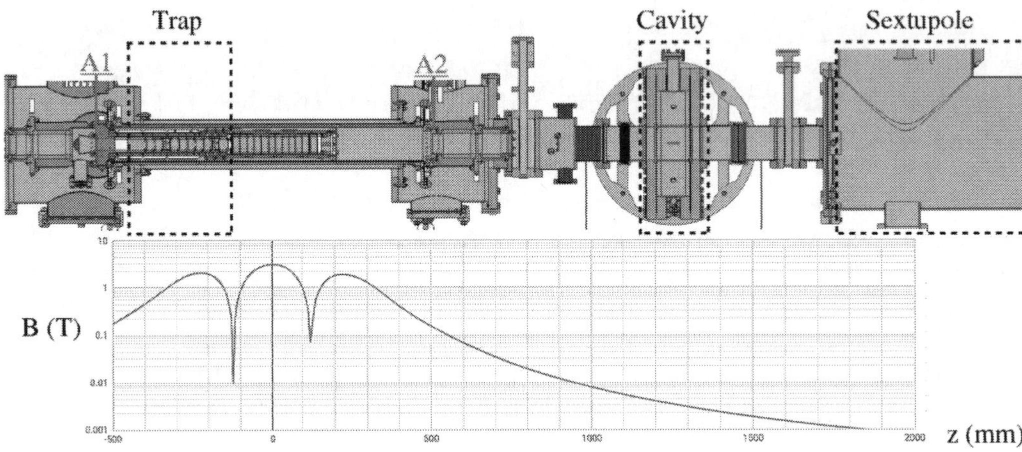

Figure 1. The Cusp experiment. Positrons and antiprotons are loaded into "Trap" from other traps further upstream (left, not shown). The antimatter plasmas are manipulated, cooled, and mixed together to form antihydrogen. The two cusps in the magnetic field B help to focus low energy, low-field-seeking antiatoms into a spin-polarized beam traveling to the right at approximately 1000 m/s. The frequency of the microwaves applied to "Cavity" is varied, while monitoring the number of antiatoms that are refocused by "Sextupole" onto a scintillator target (right, not shown). When the microwave frequency is close to the hyperfine splitting ($\nu_{HFS} \approx 1.420$ GHz), low-field-seeking atoms passing through "Cavity" become high-field-seekers. High-field-seekers are defocused by "Sextupole". Thus, the transition frequency is detected as a dip in the signal on the scintillator. "A1" and "A2" are apertures, necessary for admitting particles and letting antiatoms escape from the trap. These apertures also admit room-temperature microwave radiation, which may heat the plasma.

The number of antiatoms produced in mixing attempts so far is insufficient to perform a precision measurement; the ground state production rate must increase by about a factor of one hundred [3]. For increasing the antihydrogen yield, the most important parameter is positron temperature. A 20 K plasma should yield x100 more ground state antiatoms than the equivalent 200 K plasma [4]. In other words, the principle obstacle to this measurement appears to be the unfortunate fact that plasma in the Cusp trap tends to equilibrate closer to 200 K than to 20 K.

An electron or a positron plasma in a strong magnetic field will cool via cyclotron radiation at a rate $\Gamma = 0.26\,\mathrm{s}^{-1} \times B[\mathrm{T}]^2$ according to

$$\frac{dT}{dt} = -\Gamma(T - T_b) + H \tag{1}$$

where T is the plasma temperature, t is the time, T_b is the temperature of the thermal bath, and H is the heating rate due to plasma expansion and electrode noise. If Γ, H, and T_b are constant, T should asymptote exponentially to the value

$$T_f = T_b + H/\Gamma \tag{2}$$

If T_b is taken as the trap temperature T_t = 35 K, then a plasma temperature $T_f \sim 60$ K is a realistic expectation for typical values of H.

The factors contributing to H were minimized in the present trap by using low-noise electronics, low-pass filtering, and corrective trap alignment. It was therefore surprising to find that plasma with seemingly low H (see Section 3) relaxes to a stubbornly high temperature $T_f \approx 170\,\mathrm{K}$ (see Section 2). Much of the 2021 experimental campaign at CERN was dedicated to understanding and addressing this discrepancy.

Section 2 describes experiments using a movable thermal shield to block room-temperature radiation entering the cryogenic trap at the downstream end, equivalent to "A2" in Fig. 1. The base plasma temperature was 20 K lower when the shield was closed. This result motivated the installation of a similar shield to block the upstream end, equivalent to "A1". The mount for this shield constituted a 75% reduction in the area of "A1" and a coincident 40 K reduction in minimum plasma temperature was observed. The change in T was entirely independent of whether the small upstream shield was open or closed.

Sections 3 and 4 describe a study of expansion heating, cyclotron cooling, and final temperature as a function of the plasma's location in the trap. The study permitted a quantitative evaluation of $T_b \approx T_t + 70\,\mathrm{K}$ in the configuration with both shields installed.

In light of the accumulating evidence that the thermal bath coupling to the plasma had a room-temperature component, copper meshes were installed to completely block microwave radiation through "A1" and "A2". The resulting reduction in base plasma temperature was immediate and significant. However, those results were obtained after the EXA conference and will be reported elsewhere.

2 Movable radiation shield

A cryogenic, movable thermal shield was first used for plasma temperature studies in the ALPHA-1 experiment [5]. The magnetically actuated shield could close the entrance to the trap, preventing microwaves (and possibly residual gas) from entering. Initial observations indicated that the plasma temperature was lower when the shield was closed (J. Fajans, private communication). Povilus [6] and Carruth [7] installed similar devices in other experiments, however, without reproducing the effect on plasma temperature observed in the previous apparatus. The shield used in the present work had been used in earlier versions of the Cusp trap for achieving lower electrode temperature [8], and was removed c. 2016. It was re-installed in 2021 to see if the plasma temperature effect observed elsewhere could be reproduced in a more recent version of the Cusp trap.

Figure 1 shows the current version of the Cusp trap (winter 2021). The experiments reported here, however, were performed in an older version (summer 2021), with nearly identical electrodes but in a slightly different vacuum chamber. The downstream end of this chamber was protected from thermal radiation by the heavy copper shield mentioned above. The shield was located roughly 150 mm downstream of "A2". The shield could be opened by pulling on a braided steel cable, which is visible at the top of the photos in Fig. 2. The shield was not cryogenic because the downstream coldhead was not working at the time of these measurements. While the trap electrodes were at 35 K, the axial center of the cold bore was at 60 K and the downstream shield was at 200 K. These temperatures did not change by more than a degree over the course of weeks of experimentation, nor were they changed by opening or closing the downstream shield, whether for minutes or hours.

A plasma containing $N \approx 3 \times 10^6$ electrons was cooled under a long electrode at the magnetic field maximum. The final temperature of the plasma was found to depend on the state of the downstream thermal shield according to the middle panel of Fig. 2. A position of the shield was found for which the temperature increment associated with the shield position was roughly half that found with the shield fully open. This state, labeled "Partly Open", was photographed from several angles to judge whether it permitted a line-of-sight path into the trap. No such path existed. It is hard to imagine what could get past the "Partly Open" shield and heat the plasma other than microwaves.

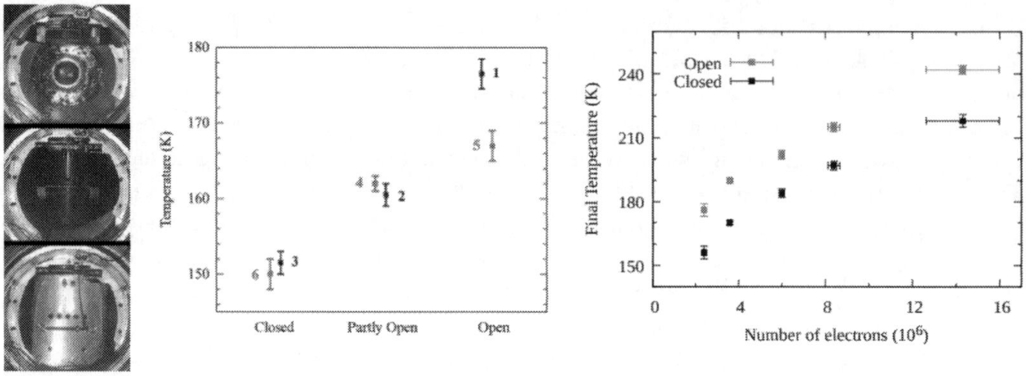

Figure 2. Left: Photos of the downstream shield taken through a vacuum window. The shield is open in the top photo, partly open in the middle photo, and closed in the bottom photo. Notice that the shield is still mostly closed in the "partly open" configuration; the trap (visible as a small circle in the top photo, having diameter roughly 1/6 the size of the photo) is still completely covered by the shield. Middle: Plasma temperature after cooling with the shield in the three positions shown at the left. Each point gives the mean and standard deviation of T for 15 different plasmas. The numbers next to the points indicate the order in which the data was taken; to further emphasize the ordering, the first three points in the series are black and the last three are red. Right: Plasma temperature as a function of the number of electrons in the plasma. The temperature is 20 K higher when the downstream shield is open, independent of the number of electrons in the plasma.

The right side of Fig. 2 shows that the temperature increment with the shield open was the same for plasmas with significantly different N (and different T_f). This provides further evidence that the "anomalous heating" has nothing to do with the plasma and may be equated to a simple offset in T_b in Eq. 2.

In response to these observations, another shield was installed at the upstream end of the trap. The temperature of the upstream shield mount was 65 K. The temperature of the electrodes remained 35 K. The shield mount reduced the upstream aperture at "A1" from 40 to 20 mm. When nearly the same sequence of plasma manipulations for generating the data in Fig. 2 was run again in this configuration, the final plasma temperature was found to be approximately 40 K lower (see Section 4). However, closing the upstream shield did not have any measurable effect on the temperature, despite much effort to ensure that the shield opened in less than 100 ms so that the plasma would not warm up while the shield was opening for the plasma temperature diagnostic (see Section 3). The remainder of this work is concerned with data obtained in this configuration, with the downstream shield closed and the upstream shield open but the aperture reduced by the shield mount.

3 Conventional plasma heating

The charges in the plasma can gain kinetic energy (plasma heating) from two principle sources. First, when the plasma expands, potential energy is directly converted into kinetic energy as charges push away from each other. Second, plasma modes can be stimulated by radiofrequency noise on the electrodes, and these modes can be damped kinetically, transferring the energy from a single plasma mode into the thermal ensemble.

Plasma expansion heating is reduced by reducing the plasma density, reducing the plasma radius, and reducing the expansion rate of the plasma. In a well-built trap with good vacuum, plasma expansion is primarily due to asymmetry transport, most often in the form of a misalignment between the magnetic field's axis of symmetry and the trap electrodes' axis of symmetry. That alignment can be roughly diagnosed by holding plasma in different axial locations (z coordinate) of the trap and slowly dumping it onto an imaging detector. If the plasma is cold and dense, most of it will come out of the radial center of the trap (at that z position), then follow the magnetic field lines to the detector. If a given magnetic field line crosses the trap center in only one of the two locations tested, then the imaged plasmas will appear in different positions on the detector. By rotating the electrode stack so as to minimize the difference between the two detected positions, the expansion rate was reduced to $1/\tau < 0.001\,\mathrm{s}^{-1}$ for most of the trap (see Fig. 3). This corresponds to a heating rate $H < 7\,\mathrm{K/s} \times nr_p^2$, where the plasma density n and radius r_p are referred to 1 mm and $1 \times 10^8\,\mathrm{cm}^{-3}$, respectively. (This heating rate is calculated as $(dU/dt)/(3/2)Nk_B$, where $dU/dt = \omega_r dP/dt$, N is the number of particles, k_B is Boltzmann's constant, ω_r is the plasma rotation rate, and P is the total canonical angular momentum of the plasma. See Eqs. (37) and (64) of Ref. [9].)

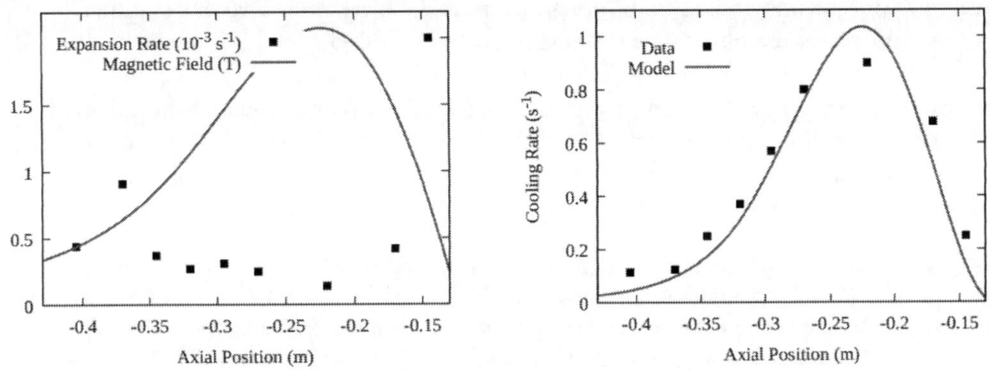

Figure 3. Left: Plasma expansion rate and axial magnetic field. Plasma containing $N \approx 3 \times 10^6$ electrons is held for $1 < t < 300\,\mathrm{s}$. The expansion rate is the reciprocal of the amount of time required for the plasma radius to double. Plasma radius is determined from the standard imaging diagnostic [10]. Right: Cooling rate compared to the formula for Γ given in Eq. 1. Cooling rate is taken as the best exponential fit to temperature vs. time data, as the plasma cools from $T \sim 6000\,\mathrm{K}$ to $T \sim 200\,\mathrm{K}$. Plasma temperature is determined from the standard temperature diagnostic [11].

Heating from electrode noise occurs when one of the normal modes of the plasma is resonant with a Fourier component of the voltage applied to the confining electrodes. Noise near the resonant frequencies of the lowest order axial modes, which are typically in the frequency range $1 < f < 50\,\mathrm{MHz}$, tends to heat the plasma the most, probably because these are the highest-Q modes and are the easiest to stimulate in the azimuthally symmetric trap geometry. These frequencies are excluded from the trap in three stages: (1) the electrode amplifiers have a bandwidth of roughly 1 kHz, (2) the filterbox mounted directly onto the air-vacuum feedthru contains single-pole lowpass filters with time constant $RC = 10\,\mu\mathrm{s}$, (3) cryogenic two-pole lowpass filters with time constant $RC = 8\,\mu\mathrm{s}$ are mounted inside the trap, 1 cm away from the electrodes. The rotating-wall electrodes (middle of the trap) and the catching electrode (upstream end of the trap) have an AC bypass to allow high frequency signals to enter. The bypass consists of an AC-coupled line on the feedthru filterboard, which is normally grounded via a relay. The signals traverse the vacuum chamber on Lakeshore Quad-Twist line, with usually two of the four twisted wires being grounded at both ends to reduce magnetic pickup. Altogether, the noise rejection seems to be very effective. A 5 V broadband noise source ($1 < f < 20\,\mathrm{MHz}$) was applied to various instruments and feedthrus entering the trap, including the MCP, the electron source, and the field ionizers (of which the wiring runs parallel to the electrode wiring internally). The only place where the noise source produced a measurable change in plasma temperature was the AC input to a rotating wall (AC bypassed) electrode, and then, only when the relays were closed. These relays are normally open except when the rotating wall is being applied.

For the trap described here, then, heating due to electrode noise seems to be negligible. Heating due to plasma expansion was calculated above to be roughly 10 K/s for most plasmas. Taken together with the cooling rate $\Gamma \gtrsim 0.4\,\mathrm{s}^{-1}$ and the trap temperature $T_t = 35$ K, one would expect plasma to cool to $T < 60$ K in the worst case, and $T < 50$ K at the maximum magnetic field. That is not what was observed.

4 Plasma temperature

Equation 2 can be used to estimate H and T_b by observing the cooling rate Γ and final temperature T_f for plasma cooled in different locations. One assumes a value for either H or T_b and solves Eq. 2 for the other. The results are given in Table 1.

The plasma used to obtain the results in Table 1 had the following parameters: $N = 3.5 \times 10^6$ electrons, $r_p = 0.7$ mm, $n = 3 \times 10^8\,\mathrm{cm}^{-3}$, $L_p \approx 4$ cm. Because the plasma length and position are different from those used to obtain the data in Fig. 3, the expansion rates were scaled by the factor $(L_p/B)^2$ [12] and averaged accordingly before applying the formula for the heating rate given in the previous section.

Table 1. For each z position, a complete T vs. t curve was taken so as to obtain experimental values for Γ and T_f. H_t is the heating rate obtained from Eq. 2 by assuming that $T_b = T_t = 35$ K. H_e is the heating rate calculated from the plasma expansion rates given in Fig. 3. T_{be} is the value of T_b obtained assuming heating is only due to expansion, i.e. setting $H = H_e$ in Eq. 2.

z (m)	Γ (s^{-1})	T_f (K)	H_t (K/s)	H_e (K/s)	T_{be} (K)
-0.38	0.09	300	24	29	-20
-0.32	0.35	130	33	13	92
-0.27	0.85	105	60	12	91
-0.22	0.90	118	75	6	112
-0.17	0.35	130	33	49	-11

The data in Table 1 leads to three conclusions.

1 The minimum plasma temperature is about 70 K higher than the expectation derived in the previous section, where it was assumed that $T_b = T_t$.

2 The higher temperatures at the two z extremes can be predicted by $T_f = H_e/\Gamma$; T_{be} is not well defined for these two cases because the relative contribution of H_e/Γ is much larger.

3 In the middle of the trap, where the cooling rate is high and the expansion rate is low, the expansion heating is entirely insufficient to account for the observed plasma temperature unless T_b is greater than T_t. The best estimate for T_{be} is about 100 K in all three cases, which is about 70 K higher than T_t. This is consistent with conclusion (1).

These observations, along with those reported in Section 2, are simply explained by the hypothesis that the temperature of the thermal bath (T_b) is generally higher than the temperature of the trap (T_t). For the purposes of cyclotron cooling, the thermal bath includes every electromagnetic mode which couples to the cyclotron motion of the plasma particles. Many of these modes are propagating, or else have a non-vanishing component outside the cryogenic region. In either case, mode energy is damped by currents on the walls of the vacuum chamber outside the cryogenic region. This raises the effective temperature of the radiation environment seen by the plasma and limits the minimum achievable plasma temperature for any trap with open ends.

Acknowledgments

This work was supported by the European Research Council under European Union's Seventh Framework Programme (FP7/2007-2013) / ERC Grant Agreement (291242); the Austrian Ministry of Science and Research, Austrian Science Fund (FWF) W1252-N27 and P 34438; the JSPS KAKENHI Fostering Joint International Research B 19KK0075; the Grant-in-Aid for Scientific Research B 20H01930; Special Research Projects for Basic Science of RIKEN; Università di Brescia and Istituto Nazionale di Fisica Nucleare; and the European Union's Horizon 2020 research and innovation programme under the Marie Skłodowska-Curie grant agreement No 721559.

References

[1] C. Malbrunot, C. Amsler, S. Arguedas Cuendis, H. Breuker, P. Dupre, M. Fleck, H. Higaki, Y. Kanai, B. Kolbinger, N. Kuroda et al., Philosophical Transactions of the Royal Society A: Mathematical, Physical and Engineering Sciences **376**, 20170273 (2018)
[2] R. Bluhm, V.A. Kosteleckỳ, N. Russell, Physical Review Letters **82**, 2254 (1999)
[3] B. Kolbinger, C. Amsler, S.A. Cuendis, H. Breuker, A. Capon, G. Costantini, P. Dupré, M. Fleck, A. Gligorova, H. Higaki et al., The European Physical Journal D **75**, 1 (2021)
[4] B. Radics, D. Murtagh, Y. Yamazaki, F. Robicheaux, Physical Review A **90**, 032704 (2014)
[5] M.J. Jenkins, *Development of an Antihydrogen Trapping Apparatus* (Swansea University, 2008)
[6] A.P. Povilus, *Cyclotron-cavity mode resonant cooling in single component electron plasmas* (University of California, Berkeley, 2015)
[7] C. Carruth, *Methods for Plasma Stabilization and Control to Improve Antihydrogen Production* (University of California, Berkeley, 2018)
[8] Y. Enomoto, *Antihydrogen production in CUSP trap* (University of Tokyo, 2011)
[9] J. Danielson, D. Dubin, R. Greaves, C. Surko, Reviews of Modern Physics **87**, 247 (2015)
[10] A. Peurrung, J. Fajans, Review of Scientific Instruments **64**, 52 (1993)
[11] B.R. Beck, *Measurement of the magnetic and temperature dependence of the electron-electron anisotropic temperature relaxation rate* (University of California, San Diego, 1990)
[12] C. Driscoll, J. Malmberg, Physical Review Letters **50**, 167 (1983)

Summary of the $K^{-}pp$ bound-state observation in E15 and future prospects

Fuminori Sakuma[1,*], *Shuhei* Ajimura[2], *Takaya* Akaishi[3], *Hidemitsu* Asano[1], *Massimiliano* Bazzi[10], *George* Beer[4], *Hyoungchan* Bhang[5], *Mario* Bragadireanu[6], *Paul* Buehler[7], *Luigi* Busso[8,9], *Michael* Cargnelli[7], *Seonho* Choi[5], *Alberto* Clozza[10], *Catalina* Curceanu[10], *Shun* Enomoto[11], *Hiroyuki* Fujioka[12], *Yuya* Fujiwara[13], *Tomokazu* Fukuda[14], *Carlo* Guaraldo[10], *Tadashi* Hashimoto[15], *Ryugo* S. Hayano[13], *Toshihiko* Hiraiwa[2], *Masami* Iio[11], *Mihai* Iliescu[10], *Kentaro* Inoue[2], *Yosuke* Ishiguro[16], *Takashi* Ishikawa[13], *Shigeru* Ishimoto[11], *Kenta* Itahashi[1], *Masahiko* Iwasaki[1], *Masaaki* Iwai[11], *Koki* Kanno[1], *Kazuma* Kato[16], *Yuko* Kato[1], *Shingo* Kawasaki[2], *Paul* Kienle[17], *Hiroshi* Kou[12], *Yue* Ma[1], *Johann* Marton[7], *Yasuyuki* Matsuda[18], *Marco* Miliucci[10], *Yutaka* Mizoi[14], *Ombretta* Morra[8], *Rie* Murayama[1], *Tomofumi* Nagae[16], *Hiroyuki* Noumi[2,11], *Hiroaki* Ohnishi[19], *Shinji* Okada[20], *Haruhiko* Outa[1], *Kyoichiro* Ozawa[11], *Kristian* Piscicchia[10,21], *Yuta* Sada[19], *Atsushi* Sakaguchi[3], *Masaharu* Sato[11], *Alessandro* Scordo[10], *Michiko* Sekimoto[11], *Hexi* Shi[7], *Kotaro* Shirotori[2], *Diana* Sirghi[6,10], *Florin* Sirghi[6,10], *Shoji* Suzuki[11], *Takatoshi* Suzuki[13], *Kiyoshi* Tanida[15], *Hideyuki* Tatsuno[22], *Makoto* Tokuda[12], *Dai* Tomono[2], *Akihisa* Toyoda[11], *Kyo* Tsukada[23], *Oton* Vázquez Doce[10,17], *Eberhard* Widmann[7], *Takumi* Yamaga[1], *Toshimitsu* Yamazaki[1,13], *Chihiro* Yoshida[19], *Qi* Zhang[1], and *Johann* Zmeskal[7]

[1]RIKEN Cluster for Pioneering Research, RIKEN, Wako 351-0198, Japan
[2]Research Center for Nuclear Physics (RCNP), Osaka University, Osaka 567-0047, Japan
[3]Department of Physics, Osaka University, Osaka 560-0043, Japan
[4]Department of Physics and Astronomy, University of Victoria, Victoria, British Columbia V8W 3P6, Canada
[5]Department of Physics, Seoul National University, Seoul 151-742, South Korea
[6]National Institute of Physics and Nuclear Engineering - IFIN HH, Bucharest, Romania
[7]Stefan-Meyer-Institut für subatomare Physik, A-1020 Vienna, Austria
[8]Istituto Nazionale di Fisica Nucleare (INFN), Sezione di Torino, Torino, Italy
[9]Dipartimento di Fisica Generale, Universita' di Torino, Torino, Italy
[10]Laboratori Nazionali di Frascati dell' INFN, I-00044 Frascati, Italy
[11]High Energy Accelerator Research Organization (KEK), Tsukuba 305-0801, Japan
[12]Department of Physics, Tokyo Institute of Technology, Tokyo 152-8551, Japan
[13]Department of Physics, The University of Tokyo, Tokyo 113-0033, Japan
[14]Laboratory of Physics, Osaka Electro-Communication University, Osaka 572-8530, Japan
[15]ASRC, Japan Atomic Energy Agency, Ibaraki 319-1195, Japan
[16]Department of Physics, Kyoto University, Kyoto 606-8502, Japan
[17]Technische Universität München, D-85748, Garching, Germany
[18]Graduate School of Arts and Sciences, The University of Tokyo, Tokyo 153-8902, Japan
[19]Research Center for Electron Photon Science (ELPH), Tohoku University, Sendai 982-0826, Japan
[20]Engineering Science Laboratory, Chubu University, Aichi 487-8501, Japan
[21]Centro Ricerche Fermi-Museo Storico della Fisica e Centro studi e ricerche "Enrico Fermi", 000184 Rome, Italy
[22]Department of Physics, Tokyo Metropolitan University, Tokyo 192-0397, Japan

*e-mail: sakuma@ribf.riken.jp

 https://doi.org/10.1051/epjconf/202226201008

[23]Institute for Chemical Research, Kyoto University, Kyoto 611-0011, Japan

Abstract. The possible existence of deeply-bound $\bar{K}$-nuclear bound states, kaonic nuclei, has been widely discussed as products of the strongly attractive $\bar{K}N$ interaction in I = 0 channels. Investigations of those exotic states will provide us unique information of the $\bar{K}N$ interaction below the threshold, which is still not fully understood so far. Recently, we observed the simplest kaonic nuclei, $\bar{K}NN$, having a much deeper binding energy than normal nuclei via in-flight (K^-, N) reactions at the J-PARC E15 experiment. For further studies, we have proposed a series of experimental programs for the systematic investigation of light kaonic nuclei, from $\bar{K}N$ ($\Lambda(1405)$) to $\bar{K}NNNN$. We will measure the $\bar{K}NNN$ ($A = 3$) system at the new experiment approved as J-PARC E80, as a first step toward a comprehensive study.

1 Introduction

The study of the $\bar{K}N$ interaction is one of the most important subjects in meson–baryon interactions in low energy quantum chromodynamics (QCD). Extensive measurements of anti-kaonic hydrogen atoms [1–3] and low-energy $\bar{K}N$ scattering [4] have revealed the strongly attractive nature of the $\bar{K}N$ interaction in the isospin I = 0 channel. Consequently, the possible existence of deeply bound kaonic nuclear states (kaonic nuclei) has been widely discussed [5–25]. Kaonic nuclei are predicted to be compact due to the strong $\bar{K}N$ attraction, suggesting that high-density nuclear matter is realized in kaonic systems.

Among kaonic nuclei, the $\bar{K}NN$ system with $I = 1/2$ and $J^P = 0^-$ (symbolically denoted as K^-pp for the $I_z = +1/2$ state) is of special interest because it is the lightest $S = -1$ $\bar{K}$ nucleus. Despite considerable experimental efforts over the past 20 years, it has been challenging to prove the existence of K^-pp. Several groups have reported observations of a K^-pp candidate with a binding energy of around 100 MeV in experiments measuring non-mesonic decay branches of Λp and/or $\Sigma^0 p$ in different reactions [26–28]. There are also contradicting reports concluding that the reactions can be understood without a bound state [29–32].

2 J-PARC E15 Experiment

We conducted an experimental investigation of the K^-pp bound state using the simplest $\bar{K}$ induced reaction of $K^- + {}^3\mathrm{He}$, via the nucleon knock-out reaction $K^-N \to \bar{K}n$ followed by the two-nucleon absorption $\bar{K} + NN \to K^-pp$. In the experiment, we used a kaon momentum of 1 GeV/c, around which the $K^-N \to \bar{K}n$ reactions have the maximum cross section. The recoiled kaon $\bar{K}$ at a momentum q behaves as an 'off-shell particle' (the total energy can be lower than its intrinsic mass) within a time range allowed by the uncertainty principle. The momentum transfer q is defined as that between the incident kaon and the outgoing neutron in the laboratory frame $q = |p_{K^-}^{lab} - p_n^{lab}|$. In the reaction, we used the low-momentum back-scattered kaon as an off-shell kaon source and the residual spectator nucleons NN as an 'actual target' to form a K^-pp state with an energy below the intrinsic mass of $M(Kpp)$ (= $m_K + 2m_N = 2.37$ GeV/c^2).

For the Λpn final states, we observed a kinematic anomaly in the Λp invariant mass near the mass threshold of $M(Kpp)$ at $q \sim 0.4$ GeV/c [33–35]. As shown in Fig. 1 (left), we confirmed the existence of the bound state below $M(Kpp)$, whose mass centroid is independent of q, at a binding energy as deep as $\sim$ 50 MeV. The back-scattered 'on-shell' kaon, whose total kaon energy is above its intrinsic mass, can also be absorbed by the spectator nucleons without forming a bound state. The kinematical centroid of this quasi-free absorption

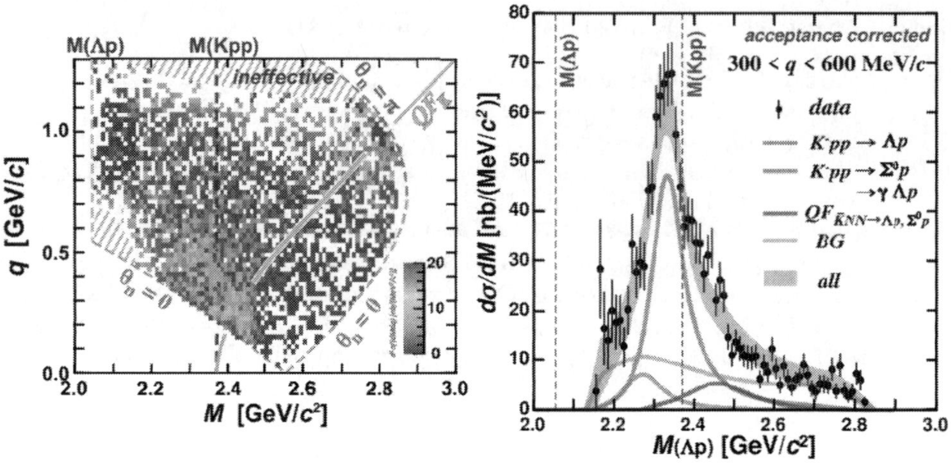

Figure 1. (left) Efficiency and acceptance corrected data over IM (Λp invariant mass) and q (momentum transfer). (right) Λp invariant mass in the region $0.3 < q < 0.6$ MeV/c.

process is plotted in the figure, denoted as QF_K. Along the line QF_K, there are two event concentration points at $\theta_n = 0$ and $\theta_n = \pi$, but both are well separated from the region of interest. Figure 1 (right) shows the Λp invariant mass spectrum for the region $0.3 < q < 0.6$ GeV/c where the K^-pp bound state is dominant. A clear peak originating from K^-pp can be seen below $M(Kpp)$, shown by the red line, whose binding energy, decay width, and S-wave Gaussian reaction form factor were observed to be $B_K = 42 \pm 3$ (stat.) $^{+3}_{-4}$ (syst.) MeV, $\Gamma_K = 100 \pm 7$ (stat.) $^{+19}_{-9}$ (syst.) MeV, and $Q_K = 383 \pm 11$ (stat.) $^{+4}_{-1}$ (syst.) MeV/c, respectively [35].

The observed binding energy of $B_K \sim 40$ MeV agrees with phenomenological predictions [10, 12, 13, 19, 36]. However, it should be noted that the obtained B_K is the spectral Breit-Wigner pole position, thus it might be different from the physical pole predicted by theoretical calculations. The observed decay width of $\Gamma_K \sim 100$ MeV is wider than the $\Lambda(1405) \rightarrow \pi\Sigma$ decay width of ~ 50 MeV (100%). If $\Lambda(1405)$ is the $\bar{K}N$ quasi-bound state, then it is naturally expected that the $\bar{K}NN \rightarrow \pi\Sigma N$ decay will occur in the same order as the $\Lambda(1405)$ decay, as is the case with most theoretical calculations. The $\bar{K}NN$ width obtained experimentally with the Λp decay in E15 is larger than that expected from the calculations, however, a possible increase of the width has been theoretically pointed out in Ref. [17] as follows. The inclusion of $\bar{K}$ absorption on the two nucleons (NN) in the $\bar{K}NN$ bound system increases the width by about 30 MeV from the YN decay to the total of about 80 MeV. This kinematical conditions for the $\bar{K}$ absorption process, $K^-pp \rightarrow \Lambda p$, would imply short-distance dynamics which is of importance to understand hadron interactions in medium, but remaining as an open question so far. The observed structure below the mass threshold in the Λp spectrum has also been theoretically interpreted as the $\bar{K}NN$ quasi-bound system: a theoretical calculation in Ref. [21, 37] demonstrated that the experimental spectrum can be reproduced with the $\bar{K}NN$ quasi-bound system and the quasi-free processes based on a theoretical treatment of the ^{3}He$(K^-, \Lambda p)n$ reaction. The observed large form factor of ~ 400 MeV/c, based on a simple plane wave impulse approximation (PWIA) [33–35], and the large binding energy of the K^-pp state would imply the formation of a compact and dense system. However, to obtain further information on the size of the $\bar{K}NN$ system, a more realistic theoretical calculation including detailed reaction dynamics and a more detailed and systematic experimental study are essential.

3 Systematic Investigation of Light Kaonic Nuclei at J-PARC

To obtain further detailed information on kaonic nuclei, we have planned a series of experimental programs using the $(K^-, N/d)$ reaction on light nuclear targets. The programs will enable a detailed study of a range of nuclei from $\bar{K}N$ ($\Lambda(1405)$) to $\bar{K}NNNN$ using the world's highest intensity low-momentum kaon beam at J-PARC. The programs comprise:

- Precise measurements of the $\Lambda(1405)$ state in a large momentum transfer region via the $d(K^-, n)$ reaction, to experimentally clarify whether it is a baryonic state or a $\bar{K}N$ molecular state,
- Investigations of the spin and parity of the $\bar{K}NN$ state via $^3\mathrm{He}(K^-, N)$ reactions,
- A search for $\bar{K}NNN$ states via $^4\mathrm{He}(K^-, N)$ reactions, as a bridge to access heavier systems, and
- An advanced search for $\bar{K}NNNN$ states via the $^6\mathrm{Li}(K^-, d)$ reaction.

In parallel to these studies, we also intend to access the $S = -2$ kaonic nuclei, such as the theoretically predicted K^-K^-pp state. The $S = -2$ system could allow us to access even higher density systems than the $S = -1$ kaonic nuclei. As described in our Letter of Intent [38], one possible approach for the measurements at J-PARC could be:

- Searching for $\bar{K}\bar{K}NN$ states via $\bar{p}^3\mathrm{He}$ annihilation.

To ensure the measurements are systematic and precise, we are planning to construct a totally new 4π spectrometer to measure all the particles involved in the reactions and to reconstruct their formation and decay exclusively. The spectrometer is designed to be highly versatile so that all the experiments can be performed simply by changing the target materials. In addition, for more efficient use of the high-intensity kaon beam, we have proposed shortening the existing K1.8BR beam line for a larger kaon yield without deteriorating the momentum resolution of the kaon beam.

4 J-PARC E80 Experiment

In the new experiment, J-PARC E80, we aim to measure the $\bar{K}NNN$ ($A = 3$) system as a first step toward a comprehensive study. The key to the experimental search is to adopt a simple reaction and to measure it exclusively. Adopting a simple reaction, such as in-flight $\bar{K}$ induced reactions with light target nuclei, enables us to specify the reaction channel using the momentum-transfer dependence. Exclusive measurements are crucial for distinguishing small and broad signals from the large and widely distributed quasi-free and multi-nucleon absorption background.

We will perform exclusive measurements of the production and decay of the K^-ppn state using the in-flight reaction

$$K^- + ^4\mathrm{He} \rightarrow K^-ppn + n$$

followed by the expected no-mesonic decays

$$\begin{aligned} K^-ppn &\rightarrow \Lambda + d, \\ K^-ppn &\rightarrow \Lambda + p + n. \end{aligned}$$

We aim to determine the binding energy and width from the invariant mass reconstruction of the decays. The invariant mass will be obtained as a function of the momentum transfer to distinguish the bound-state production from the quasi-free processes and multi-nucleon absorption processes by the event kinematics as demonstrated in the E15 analysis. In the

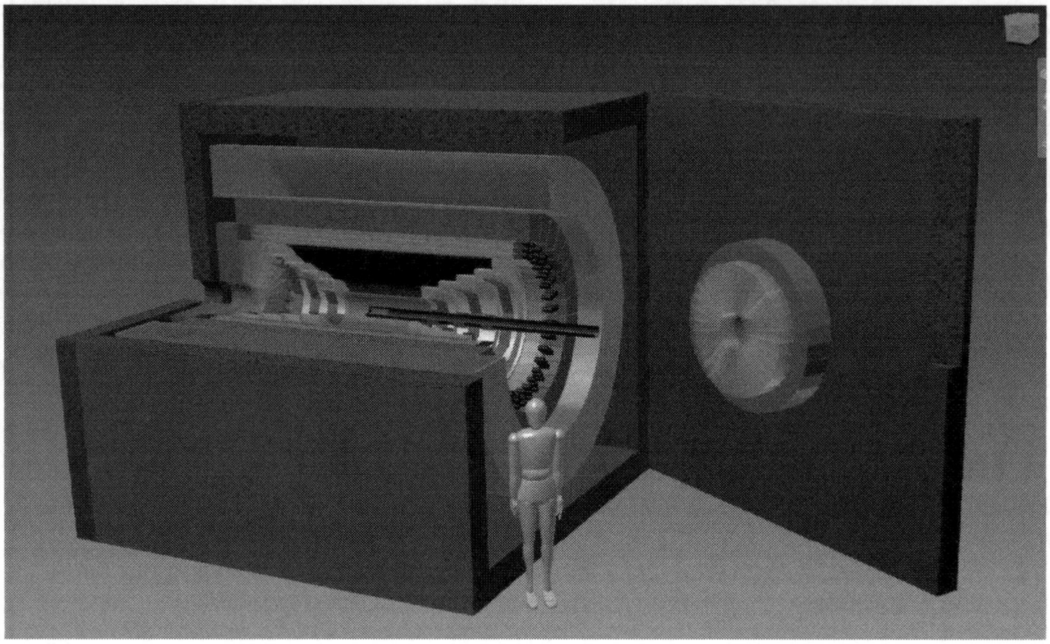

Figure 2. Conceptual design of the CDS.

$K^- + {}^4$He reaction, it is possible to measure the isospin partner of K^-ppn, *i.e.*, the K^-pnn state via

$$\begin{aligned} K^- +^4 \mathrm{He} &\rightarrow K^-pnn + p, \\ K^-pnn &\rightarrow \Lambda + n + n. \end{aligned}$$

Comparing the properties of the isospin partners is of special importance for investigating the internal composition of the kaonic nuclei. On the other hand, the measurement is challenging, because two-neutron detection is required to identify the K^-pnn decay.

We will utilize 1.0 GeV/c incident kaons to maximize the $\bar{K}N$ reaction rate at around zero degrees. The incoming K^- beam will be identified and its momentum analyzed by the beam-line spectrometer. The beam kaon will irradiate a liquid ^{4}He target located at the final focus point, and all the particles generated from the reactions will be identified with a newly constructed cylindrical detector system (CDS) surrounding the target system. A conceptual design of the CDS is shown in Fig. 2. It is mainly composed of a large superconducting solenoid magnet, a cylindrical wire drift chamber, and a cylindrical neutron detector. The kaonic nuclei will then be identified via invariant-mass reconstruction of the decay particles. By detecting the nucleon coming from the initial (K^-, N) reaction, or by identifying it with the missing mass technique, we will realize exclusive measurement of the kaonic nuclei. The details of the apparatus used for the experiment can be found in Ref. [39].

The new experiment will provide the mass number dependence of the kaonic nuclei for the first time. The dependence can more clearly reveal the $\bar{K}N$ interaction below the mass threshold, by comparing the obtained properties of the $\bar{K}NNN$ state with those of the already reported K^-p ($\Lambda(1405)$) and K^-pp states.

5 Summary

We demonstrated that kaonic nuclei can be produced via in-flight (K^-, N) reactions using the low-momentum DC kaon beam at the J-PARC E15 experiment. We observed that the simplest kaonic nuclei, K^-pp, has a much deeper binding energy than normal nuclei. We also found that the large form factor obtained in a PWIA analysis implies the possible formation of a compact and dense system. For the next stage, we have proposed a series of experimental programs for the systematic investigation of light kaonic nuclei, from $\bar{K}N$ ($\Lambda(1405)$) to $\bar{K}NNNN$. Through the experiments, we will determine the features of kaonic nuclei depending on the mass number A, *i.e.*, nuclear density, which is related to spontaneous and explicit chiral symmetry breaking in QCD. In the new experiment approved as J-PARC E80 we will measure the $\bar{K}NNN$ ($A = 3$) system as a first step toward a comprehensive study.

6 Acknowledgements

We are grateful to all the staff members of J-PARC/KEK/JAEA for their extensive efforts in the successful operation of the facility. This work is supported in part by MEXT Grants-in-Aid 26800158, 17K05481, 26287057, 24105003, 14102005, 17070007, 18H05402, 18H01237, and 20K04006. Part of this work is also supported by the Ministero degli Affari Esteri e della Cooperazione Internazionale, Direzione Generale per la Promozione del Sistema Paese (MAECI), StrangeMatter project.

References

[1] M. Iwasaki et al., Phys. Rev. Lett. **78**, 3067 (1997)
[2] G. Beer et al., Phys. Rev. Lett. **94**, 212302 (2005)
[3] M. Bazzi et al., Phys. Lett. **B704**, 113 (2011)
[4] A.D. Martin, Nucl. Phys. **B179**, 33 (1981)
[5] Y. Nogami, Phys. Lett. **7**, 288 (1963)
[6] Y. Akaishi, T. Yamazaki, Phys. Rev. **C65**, 044005 (2002)
[7] T. Yamazaki, Y. Akaishi, Phys. Lett. **B535**, 70 (2002)
[8] N.V. Shevchenko, A. Gal, J. Mares, Phys. Rev. Lett. **98**, 082301 (2007)
[9] N.V. Shevchenko, A. Gal, J. Mares, J. Révai, Phys. Rev. **C76**, 044004 (2007)
[10] Y. Ikeda, T. Sato, Phys. Rev. **C76**, 035203 (2007)
[11] A. Doté, T. Hyodo, W. Weise, Nucl. Phys. **A804**, 197 (2008)
[12] Y. Ikeda, T. Sato, Phys. Rev. **C79**, 035201 (2009)
[13] S. Wycech, A.M. Green, Phys. Rev. **C79**, 014001 (2009)
[14] A. Doté, T. Hyodo, W. Weise, Phys. Rev. **C79**, 014003 (2009)
[15] Y. Ikeda, H. Kamano, T. Sato, Prog. Theor. Phys. **124**, 533 (2010)
[16] N. Barnea, A. Gal, E.Z. Liverts, Phys. Lett. **B712**, 132 (2012)
[17] M. Bayar, E. Oset, Phys. Rev. **C88**, 044003 (2013)
[18] S. Maeda, Y. Akaishi, T. Yamazaki, Proc. Jpn. Acad. **B89**, 418 (2013)
[19] J. Révai, N.V. Shevchenko, Phys. Rev. **C90**, 034004 (2014)
[20] A. Doté, T. Inoue, T. Myo, Prog. Theor. Exp. Phys. **2015**, 043D02 (2015)
[21] T. Sekihara, E. Oset, A. Ramos, Prog. Theor. Exp. Phys. **2016**, 123D03 (2016)
[22] S. Ohnishi, W. Horiuchi, T. Hoshino, K. Miyahara, T. Hyodo, Phys. Rev. **C95**, 065202 (2017)
[23] A. Doté, T. Inoue, T. Myo, Phys. Rev. **C95**, 062201 (2017)

[24] A. Doté, T. Inoue, T. Myo, Phys. Lett. **B784**, 405 (2018)
[25] Y. Kanada-En'yo, Eur. Phys. J. A **57**, 185 (2021)
[26] M. Agnello et al., Phys. Rev. Lett. **94**, 212303 (2005)
[27] T. Yamazaki et al., Phys. Rev. Lett. **104**, 132502 (2010)
[28] Y. Ichikawa et al., Prog. Theor. Exp. Phys. **2015**, 021D01 (2015)
[29] O. Vázquez Doce et al., Phys. Lett. **B758**, 134 (2016)
[30] R. Del Grande et al., Eur. Phys. J. **C79**, 190 (2019)
[31] G. Agakishiev et al., Phys. Lett. **B742**, 242 (2015)
[32] A.O. Tokiyasu et al., Phys. Lett. **B728**, 616 (2014)
[33] Y. Sada et al., Prog. Theor. Exp. Phys. **2016**, 051D01 (2016)
[34] S. Ajimura et al., Phys. Lett. **B789**, 620 (2019)
[35] T. Yamaga et al., Phys. Rev. C **102**, 044002 (2020)
[36] T. Yamazaki, Y. Akaishi, Phys. Rev. **C76**, 045201 (2007)
[37] T. Sekihara, E. Oset, A. Ramos, JPS Conf. Proc. **26**, 023009 (2019)
[38] Letter of Intent for J-PARC, "Double Anti-kaon Production in Nuclei by Stopped Anti-proton Annihilation" (2009), http://j-parc.jp/researcher/Hadron/en/Proposal_e.html
[39] Proposal for J-PARC, "Systematic investigation of the light kaonic nuclei: via the in-flight ^{4}He(K^-, N) reactions" (2020), http://j-parc.jp/researcher/Hadron/en/Proposal_e.html

Recent results on the positronium decay studies with the J-PET detector

Eryk Czerwiński[1,*] and *Juhi* Raj[1] for the J-PET Collaboration

[1]Marian Smoluchowski Institute of Physics, Jagiellonian University, Poland

Abstract. Positronium, as a bound state of electron and positron and the lightest matter-antimatter system and at the same time an eigenstate of the C and P operators is a unique probe to search for possible violation of combined charge, parity, and time-reversal symmetries (CPT). The test is performed by a measurement of angular correlations in the annihilations of the lightest leptonic bound system. The J-PET detector is the only device which enables the determination of the polarization of photons from positronium annihilation together with the positronium spin axis on an event-by-event basis. This allows to explore a new class of discrete symmetry odd operators that were not investigated before. The first test of CPT symmetry at J-PET is presented together with preliminary results of CP, P and T symmetry test.

1 Introduction

Motivations to study CP and CPT invariance are manifold [1–3]. One of them being that the CPT theorem reveals foundational character of CPT symmetry conservation requirement, which is equivalent to usage of unitary, local and Lorentz-invariant quantum field theory. As such, tests of CPT invariance probe fundamentals of physics [4]. Moreover, CPT contains charge conjugation, and therefore represents a symmetry between matter and antimatter. The amount of CP violation contained in the Standard Model appears to be insufficient for a convincing explanation of the observed prevalence of matter over antimatter in the Universe [5, 6]. For these reasons, discrete symmetry tests remain an interesting experimental research area in fundamental physics. One possible way to study the discrete symmetry violation is to investigate the expectation values for the symmetry odd-operators of the non-degenerate stationary states. The anti-unitary character of the discrete-symmetry operator makes the experimental studies of the time reversal invariance more challenging than other symmetries, since it requires abilities of preparing the initial and final states of the process in a fully controlled way [7]. So far, none of the experiments reported limits on the T-symmetry violations in the decays of positronium. The known final state interactions of photons are expected to mimic the T-symmetry violation at the level of 10^{-9} [8, 9]. All of the previous investigations with positronium, which tested the discrete symmetries odd operators, were based on the products of photons momenta and positronium spin vectors [10]. The J-PET group extends the study to other operators, taking advantages of properties of the J-PET tomography scanner, which enables to determine the momentum direction of the secondary

*e-mail: eryk.czerwinski@uj.edu.pl

scattered photons [7]. Using the J-PET detector, the discrete-symmetries can be investigated by searching for the possible non-zero expectation values of the symmetry-odd operators which are constructed from the momentum vectors of the primary and secondary scattered photons originating from the decays of the o-Ps atoms [7].

2 Jagiellonian - Positron Emission Tomograph

PET (positron emission tomography) is a nuclear medicine functional imaging technology for observing metabolic processes in the body as a tool for tumor diagnostics at the cellular level [11–14]. For the detection of annihilation photons, all commercially available PET scanners use relatively expensive crystal detectors. The Jagiellonian-Positron Emission Tomograph (J-PET) is the first PET scanner to use plastic scintillator strips, making it more cost efficient and portable [15–18]. The J-PET detector's ability to measure the polarization of annihilation photons is one of its distinguishing properties [7, 19, 20]. The detector is made up of 192 plastic scintillator strips (EJ-230) with dimensions of $500 \times 19 \times 7\text{mm}^3$ each that are arranged in three concentric layers (48 modules on radius 425 mm, 48 modules on radius 467.5 mm, and 96 modules on radius 575 mm). Each scintillator of the J-PET scanner is optically coupled to Hamamatsu R9800 vacuum tube photomultipliers on both ends [21–23]. This yields 384 analog channels, which are processed by a fully equipped trigger-less Data Acquisition System (DAQ) and readout mechanism [24–26].

The experiments were conducted with a point-like ^{22}Na source placed in the center of the detector and surrounded by XAD-4 porous polymer to produce positronium atoms [27]. The positrons emitted from the source ($^{22}Na \rightarrow {}^{22}Ne^* + e^+ + \nu_e$) combine with the electrons in the XAD-4 porous polymer to produce ortho-Positronium (o-Ps), a meta-stable triplet state that primarily decays into three photons due to charge conjugation symmetry conservation.

The key feature of the J-PET detector is its fast timing properties from plastic scintillators which allows the calculation of the relative azimuthal angles between the interacting photons. From kinematics we infer that the sum of the two smallest relative azimuthal angles between the registered annihilation photons for $o\text{-}Ps \rightarrow 3\gamma$ must be greater than 180° [7]. Using this feature of the detector we are capable of suppressing a large sample of background of e+e- $\rightarrow 2\gamma$.

3 Results

Recently the J-PET collaboration reported a measurement of the expectation value of the angular correlation $O_{CPT} = \frac{\vec{S}\cdot(\vec{k_1}\times\vec{k_2})}{|\vec{k_1}\times\vec{k_2}|}$, where $\vec{S}$ denotes the polarization vector of o-Ps and $\vec{k_i}$ is the momentum of the i-th annihilation photon [28]. With a 10 MBq sodium source a sample of almost 10^7 event candidates from the o-Ps$\rightarrow 3\gamma$ was collected. Since parity is violated in the β decay, the emitted positrons are polarized longitudinally along the momentum direction [7]. In the production of Ps this polarization is preserved to a large degree [29]. Therefore, the polarization of the o-Ps can be estimated from the emission point of the positron and annihilation point of o-Ps. The sensitivity of O_{CPT} reached by J-PET is three times better than the previous result [28]. The error has a negligible systematical contribution and the result of measurement shows no CPT symmetry violation.

Based on 122 days of data taking with 1 MBq source we report preliminary accuracy of the measurement of the expectation value of the following operator: $O_{CP,P,T} = \frac{\vec{k_i}\cdot(\vec{k_j}\times\vec{k'_j})}{|\vec{k_j}\times\vec{k'_j}|}$. Here $\vec{k_k}$ denotes the momentum of the k-th annihilation photon and $\vec{k'_k}$ is the momentum of the scattered photon [7, 30]. With the available statistics of 7×10^5 events we are able to reach an accuracy of 10^{-4} without contribution from systematical uncertainties. It is worth to mention that the previous test of CP symmetry with o-Ps reached the level of 10^{-3} [10].

4 Conclusion

First measurement of expectation value of symmetry-odd operator reported by J-PET group, together with prospects of upcoming results, confirms the potential of the detector for tests of discrete symmetries [28, 31]. This device has unique property of combining determination of photon polarization together with estimation of positronium spin axis on the event-by-event basis. Although the accuracy of CPT and CP, P, T tests are statistically dominated, the J-PET detector undergoes an upgrade for an additional layer of scintillators read out by silicon photomultipliers (SiPMs), which would allow to triple the efficiency for the single photon detection and improve the time resolution by about a factor of 1.5.

5 Acknowledgements

The J-PET collaboration acknowledges the support provided by the Polish NationalCenter for Research and Development through grant INNOTECH-K1/IN1/64/159174/NCBR/12; the Foundation for Polish Science through the MPD and TEAMPOIR.04.04.00-00-4204/17 Programmes; the National Science Centre of Polandthrough grant nos. 2016/21/B/ST2/01222, 2017/25/N/NZ1/00861, and 2019/35/B/ST2/03562; the Ministry for Science and Higher Education through grant nos. 6673/IA/SP/2016 and 7150/E- 338/SPUB/2017/1; the Jagiellonian University via project no.CRP/0641.221.2020, as well as the SciMatand DigiWorld Priority Research Area budgets under the programme ExcellenceInitiative - Research University at the Jagiellonian University.

References

[1] E. Noether, Mathematisch-Physikalische Klasse, **5591918** 235–257 (1918).
[2] J. H. Christenson, J. W. Cronin, V. L. Fitch, and R. Turlay, Phys. Rev. Lett., **13**, 138–140 (1964).
[3] S. Bass, Acta Phys. Pol., B**50** No. 7:1319 (2019).
[4] G. S. Adkins, R. N. Fell, and J. Sapirstein. Ann. Phys., **295**, 136 (2002).
[5] A. Pokraka and A. Czarnecki, Phys. Rev. D96, 093002 (2017).
[6] P. A. Vetter and S.J. Freedman, Phys. Rev. Lett. 91, 263401 (2003).
[7] P. Moskal et al., Acta Phys. Polon. **B 47**, 509 (2016).
[8] W. Bernreuther and O. Nachtmann. Z. Phys., C **11**, 235 (1981).
[9] B. K. Arbic et. al., Phys. Rev. A 37, 3189 (1988).
[10] T. Yamazaki et al., Phys. Rev. Lett. **104**, 083401 (2010).
[11] P. Moskal and E. Stepien, PET Clin, **15**,439-452 (2020).
[12] P. Moskal et al. Phys. Med. Biol., **66** 175015 (2021).
[13] P. Moskal, P. Jasinska, E. Stepien, and S. Bass. Nature Reviews Physics, **1**, 527–529 (2019).
[14] P. Moskal and K. Dulski et al., Sci. Adv., **7**, Issue 42:eabh4394 (2021).
[15] R. Shopa et al. Medical Image Analysis, **73** 102199, (2021).
[16] S. Niedźwiecki, et al., Acta Phys. Pol. **B 48**, 1567 (2017).
[17] L. Kaplon et al., Bio-Algorithms and Med-Systems. Volume **10**, Issue 1, Pages 27-31 (2014).

[18] P. Moskal et al., IEEE Transactions on Instrumentation and Measurement **70**, 2000810 (2021).
[19] P. Moskal and N. Krawczyk et al. Eur. Phys. J. C, **78**, 970 (2018).
[20] B. C. Hiesmayr and P. Moskal. Scientific Reports, **9**,8166 (2019).
[21] P. Moskal et al., Nucl. Instr. and Meth. A **775**, 54 (2015).
[22] L. Raczynski et al. Nucl. Instrum. Methods, A **786**, 105 (2015).
[23] L. Raczyński et al. Phys. Med. Biol., **62**, 5076 (2017).
[24] G. Korcyl et al. Acta Phys. Pol., B **47**, 491 (2016).
[25] G. Korcyl et al. IEEE Transactions on Medical Imaging Vol. **37**, No. 11, 2526 (2018).
[26] W. Krzemień et al. SoftwareX, **11**, 100487 (2020).
[27] B. Jasinska et al., Acta Phys. Polon., B **47**, 453 (2016).
[28] P. Moskal and A. Gajos et al. Nature Commun., **12**, 5658 (2021).
[29] P. Coleman, Positron Beams and Their Applications, World Scientific (2000).
[30] J. Raj, K. Dulski, and E. Czerwinski. Acta Phys. Pol., B 51:149 (2020).
[31] A. Gajos et al. Advances in High Energy Physics, vol. **2018** (2018).

A search for atomic parity violation in muonic atoms using a high-intensity pulsed muon beam at J-PARC

Sohtaro Kanda[1,*]

[1]Institute of Materials Structure Science, KEK, 1-1 Oho, Tsukuba, Ibaraki 305-0801, Japan

Abstract. An exotic atom consisting of a negative muon and a nucleus is a muonic atom. Atomic parity violation in muonic atoms provides a unique opportunity to determine the Weinberg angle and search for physics beyond the standard model at the low energy scale. We have proposed a new experiment to perform X-ray spectroscopy of muonic atoms using a high-intensity pulsed muon beam at J-PARC. A scintillator-based calorimeter and a low-density gaseous target were developed for the experiment. As a feasibility study, we measured the yield of muon atoms and searched for muonic atom's metastability using low-density methane gas. An overview of the experiment and the test experiment results are presented.

1 Introduction

In the standard model of particle physics, the Weinberg angle is an energy-dependent parameter that describes the mixing of the electromagnetic and weak interactions [1, 2]. The modified minimal subtraction scheme predicts the scale dependence of the Weinberg angle precisely [3, 4]. Measurements of the Weinberg angle at various energy scales are essentially important as a precision test of the standard model and search for new physics. The Weinberg angle can be determined via the neutral current interactions, mediated by the Z boson. At the low energy scale, where the typical momentum transfer is on the order of MeV [5], measurements of atomic parity violation (APV) give a precise value of the Weinberg angle. Atomic parity violation refers to a parity non-conservation (PNC) effect in atomic systems induced by the interference between the electromagnetic and weak interactions. In the leading order, the Weinberg angle θ_W and the weak charge Q_W obey the following equation,

$$Q_W = -(N - Z + 4Z\sin^2\theta_W), \tag{1}$$

where N and Z are the neutron and atomic numbers, respectively.

In APV experiments, the parity-violating asymmetry A_{LR} is obtained as the magnitude of the left-right asymmetry with respect to a particular direction. A naive estimation of the asymmetry's magnitude is

$$A_{LR} \sim \alpha^2 \frac{m_e^2}{m_Z^2} \sim 10^{-15}, \tag{2}$$

which is so small that observation is impractical. However, the enhancement mechanism called the Z^3 law makes the magnitude observable to the order of 10^{-6}. Heavy atoms are

*e-mail: kanda@post.kek.jp

advantageous for observing the asymmetry, and experiments have been performed on cesium [6, 7], ytterbium [8], thallium [9, 10], lead [11], and bismuth [12]. The most precise result to date was obtained by the experiment using cesium atom [6]. New experiments using francium [13] and radium [14] are in progress.

According to Eq. (2), the magnitude of APV is proportional to the squared mass of the lepton used as a probe. An APV measurement using muons can be a powerful method to determine the weak charge and the Weinberg angle.

When a nuclear Coulomb potential captures a negative muon, electrons around the nuclei are kicked out, and the muon forms an exotic bound state called a muonic atom. The muon in orbit is highly sensitive to the interactions between the nucleus since the muon is 200 times heavier than the electron. An APV experiment using muonic atoms provides a unique opportunity to search for physics beyond the standard model. In particular, the models predicting lepton-universality breaking are important candidates to be tested [15].

Although APV measurements using muons have been proposed theoretically since the 1970s [16, 17], there were various technical difficulties at that time. In the 1990s, several experiments were carried out at Paul Scherrer Institute (PSI) using a cyclotron trap [18, 19]. The experiments have provided valuable insights into the de-excitation dynamics of muonic atoms. However, no parity-odd transition was observed. Recently, a new APV experiment using a mixture of krypton in hydrogen has been proposed at PSI, and feasibility studies are underway [20].

At J-PARC MLF MUSE [1], a high-intensity low-energy negative muon beam is available. This is beneficial to obtain muonic atoms in a low-density gaseous target without a cyclotron trap. Towards the first observation of APV in muonic atoms, a new experiment using the high-intensity pulsed muon beam and a segmented calorimeter with fast signal processing is proposed.

2 Atomic parity violation in muonic atoms

Muonic atoms form in excited states and then sequentially de-excite through the cascade processes such as radiative and Auger transitions and Coulomb de-excitation. Typically, the principal quantum number n in an initial state is $\sqrt{m_\mu/m_e} \sim 14$ where m_μ and m_e are the muon and electron masses, respectively. Most of the muons in atoms de-excite to the muonic ground-state immediately after formation. However, for a case of a low-density gaseous target with a small atomic number, several percent of the muons stay in metastable states with $n = 2$.

The process of interest is the 2S-1S transition with one-photon emission. The selection rule suppresses this process. However, PNC mixing between 2S and 2P states enhances the probability of the one-photon transition. The transition is parity-odd, and X-ray's emission angle is anisotropic like an electron from parity-violating muon decay.

Figure 1 summarizes the APV-relevant atomic properties as a function of the atomic number Z. The asymmetry increases with Z, but the signal-to-noise ratio S/N behaves oppositely. The S/N ratio can be defined as the ratio of the rate of one-photon transition to the rate of the others. It is tiny, at a level of sub-percent at most, so a technique for background rejection is essential. For small Z, the difference in lifetime between 2S and 2P separates the signal from backgrounds, and for large Z, the energy difference can be used for background rejection. For the experiment, carbon (Z=6) is selected as the target, considering the asymmetry, S/N, and metastability's lifetime.

[1] MLF and MUSE stand for Materials and Life Science Experimental Facility and Muon Science Establishment, respectively.

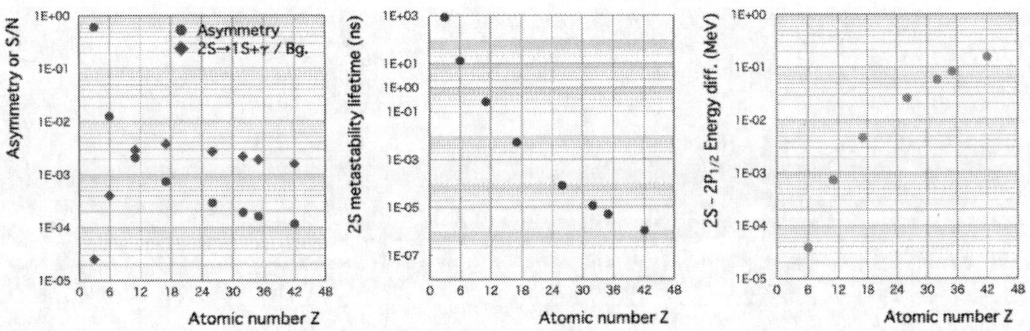

Figure 1. Atomic number dependence of some physical quantities relevant to APV measurements: (left) the APV asymmetry and the signal-to-noise ratio, (center) the lifetime of 2S metastability, (right) the energy difference between 2P and 2S states. The values are from tables I, II, and III in the reference [16].

3 Experimental method

This section outlines the experimental design, estimates the statistical sensitivity by simulation, and describes the detector development.

3.1 Overview

Figure 2 illustrates the experimental apparatus to observe the angular asymmetry of X-rays from the parity-violating transition. A polarized negative muon beam irradiates a low-density gas target. The muons form atomic bound states after slowing down by ionization. The segmented calorimeter consisting of Ce:LYSO[2] crystals and silicon photomultipliers (SiPMs) detect the X-rays from muonic atoms. A fiber hodoscope for electron tracking measures the muon stopping distribution in the target and optimizes the beam energy.

3.2 Simulation

The statistical precision of the experiment was evaluated by Monte-Carlo simulations considering the muon stopping distribution, rates of transitions, and energy distribution of emitted photons. Figure 3 shows simulation results assuming the energy resolution of 10% (FWHM) and the beam intensity of 5×10^5 Hz. Even if there is no enhancement by new physics, we expect to observe statistically significant asymmetry in two weeks of measurement.

3.3 Detector

The calorimeter unit consists of 16 SiPMs (Hamamatsu Photonics, MPPC S14160-6050HS) in a 4-by-4 arrangement. The bias voltage for SiPMs is applied in parallel, and the signal line is connected in series to be read out as a single channel. The signal is recorded by a waveform digitizer (CAEN, DT5730S) via an amplifier (Kaizuworks, KN2107). A 2.54 cm square, 3 mm thick LYSO:Ce scintillator is placed on top of the SiPM array and covered with aluminized mylar thin film. The detector design was determined based on the results of a test experiment conducted at the RIKEN-RAL muon facility in 2019 [21].

[2]Ce doped $Lu_{1.9}Y_{0.1}SiO_5$

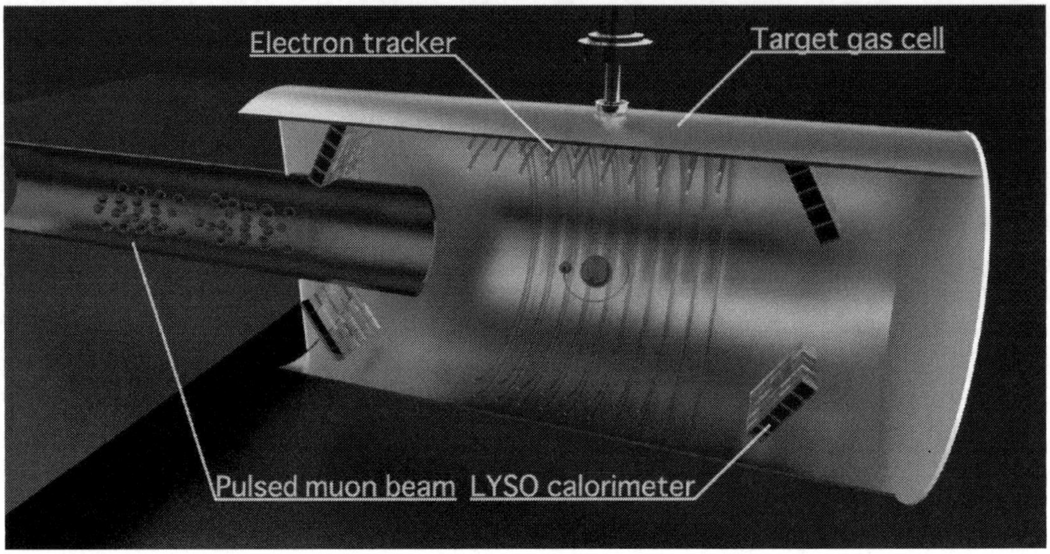

Figure 2. Experimental schematic. A low-pressure gas target is confined in an aluminum cell. A calorimeter using an inorganic scintillator and an electron tracker using scintillating fibers are placed in the cell to detect X-rays and electrons. The track reconstruction of decay electrons determines the spatial distribution of muonic atoms, and the effect of APV is quantified from the forward/backward asymmetry of muonic X-rays with respect to the muon spin direction.

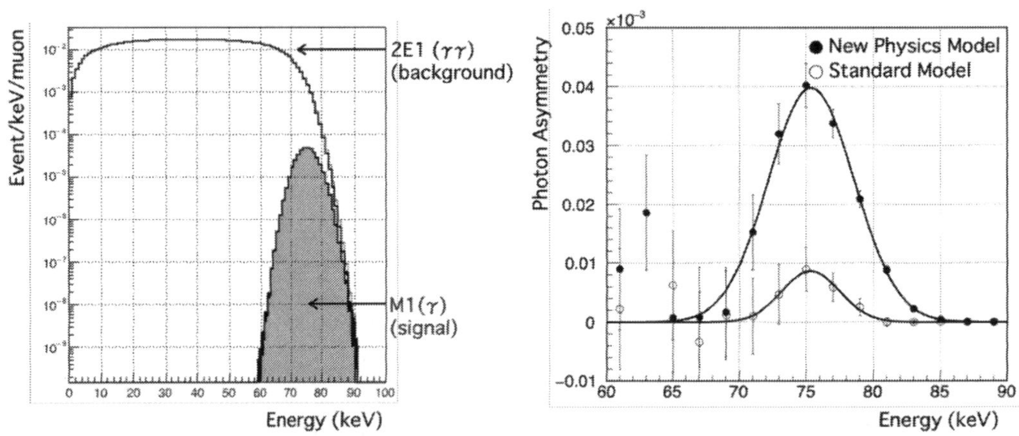

Figure 3. Results of a simulation study: (left) the energy spectra of photons from one- and two-photon transitions assuming the energy resolution of 10% FWHM, (right) the forward/backward asymmetry of photons. The enhancement of the asymmetry by the new physics was assumed to be the maximum value in the reference [15].

4 Results of a feasibility study

A feasibility study using a high-intensity pulsed muon beam was performed at the J-PARC MLF MUSE D-Line in April 2020. The purpose of this experiment is to optimize the muon beam momentum and to search for the 2S metastability of muonic carbon atoms with a low-

Figure 4. Photographs of the apparatus: (left) the SiPM array for calorimeter, (right) the gas cell with the calorimeters. The SiPM array consists of 16 SiPMs with a total active area of 2.54 cm × 2.54 cm. The LYSO:Ce scintillator is placed on the array (not shown in the left figure).

density methane gas target. The target cell was placed inside the μSR spectrometer [22]. The spectrometer consists of electron detectors and coils for the longitudinal and transverse fields.

We succeeded in detecting X-rays with 75 keV energy from muonic carbon atoms using a methane gas target at 0.3 atm. Detailed analysis to search for 2S metastability is in progress. In this section, we will discuss the analysis of decay electrons to evaluate the muonic atom yields and a search for muonic carbon atoms in the 2S metastable state.

Typical time spectra of decay electrons are shown on the left side of Fig. 5. The spectra were analyzed with four components taking into account the lifetime of muonic atoms to evaluate the yield of each component. Carbon and hydrogen correspond to the target gas, aluminum to the target cell, and lead to the beam collimator.

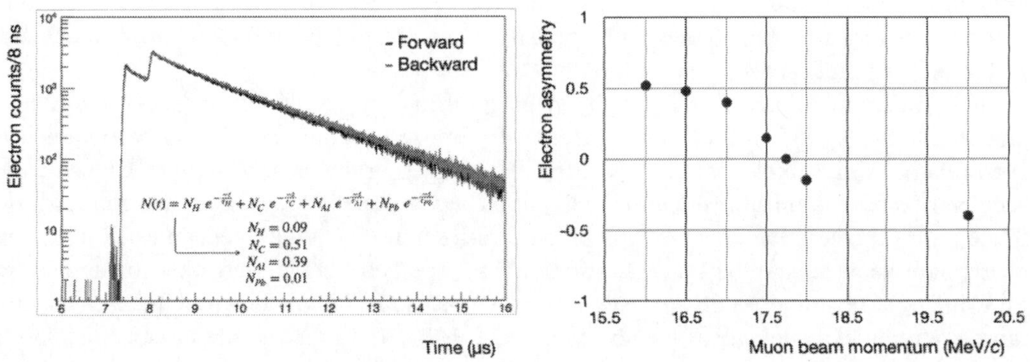

Figure 5. Experimental results: (left) decay electron time spectra obtained by the μSR spectrometer with the beam momentum of 17.75 MeV/c at the target pressure of 0.3 atm, (right) forward/backward asymmetry of decay electrons. Zero-asymmetry means that the beam stops in the center of the gas target.

The beam's momentum is optimized so that the center of the muon stopping distribution coincides with the target cell center. When this is realized, the count of electrons originating from muonic carbon atoms is equal forward and backward. The decay electron's forward/backward asymmetry is shown on the right side of Fig. 5. At a target pressure of 0.3 atm, the asymmetry was almost zero at the momentum of 17.75 MeV/c.

After momentum optimization, the 2S metastability of muonic carbon atoms was searched for. The observation of APV requires fully-ionized muonic carbon. If they are produced, muon spin rotation can be observed in a transverse magnetic field. Figure 6 shows analysis results. No significant rotation was observed in the experiment. On the other hand, spin relaxation was observed in the longitudinal field. This result suggests the existence of residual electrons, but further investigation is necessary because it may be due to other causes such as changes in muon stopping distribution or contributions from muons that have stopped beside the gas target.

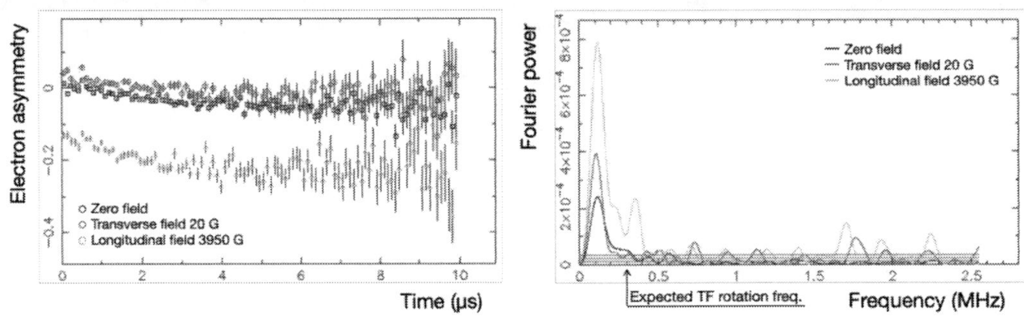

Figure 6. Results of decay electron analysis: (left) the forward/backward asymmetry as a function of time, (right) the Fourier power spectra obtained from the time spectra.

For the case of the 0.3 atm target, there is some ambiguity as to whether the muonic atom can be regarded as an isolated system, and the possibility of electron refilling from surrounding atoms cannot be ruled out. In this experiment, we also confirmed the production of muonic carbon atoms at 0.1 atm. However, the muon stopping efficiency was low, and it was difficult to obtain sufficient statistics in the limited beam time. In order to make experiments at lower gas pressures feasible, we are working on increasing the solid angle of the detector and improving the signal-to-noise ratio.

It is important to clarify what is happening during the formation and de-excitation of muonic atoms. In the case of noble gases, fully-ionized muonic atoms can be obtained at even relatively high pressures [23]. On the other hand, in the case of molecular gases, the acceleration of muonic atoms due to the Coulomb explosion causes further electron refilling. Since the re-acceleration energy depends on the structure of the gas molecules [24], targets other than methane may be the best solution. Ethane and butane were investigated in the experiment at PSI, and no 2S metastability was observed [25]. It is worthwhile to explore more possibilities systematically with numerical calculations of a cascade model considering the Coulomb explosion.

5 Summary

Atomic parity violation in muonic systems opens a unique portal to study new physics, but experiments have been challenging. A new project was proposed using a high-intensity pulsed muon beam and a fast scintillator-based calorimeter. The first experiment was performed at J-PARC MLF as a feasibility test, and the beam momentum was optimized. The proposal for a continuing experiment has been approved, and it will be conducted in February 2022.

6 Acknowledgements

The author thanks the J-PARC and KEK personnel for scientific, engineering, and technical supports to prepare and perform the experiment. In particular, we would like to thank Dr. Patrick Strasser for his great help in handling the gas target. The experiment was conducted under the user program of 2020B0092 at MLF J-PARC. This work was supported by the Grant-in-Aid for Scientific Research from JSPS/MEXT KAKENHI Grant Numbers 19H04618 and 21H00172.

References

[1] S. L. Glashow, Nucl. Phys. **22**, 579 (1961).
[2] S. Weinberg, Phys. Rev. Lett. **19**, 1264 (1967);
[3] J. Erler and M. J. Ramsey-Musolf, Phys. Rev. D **72**, 073003 (2005).
[4] J. Erler and R. Fernández, JHEP **1803**, 196 (2018).
[5] C. Bouchiat and C. A. Piketty, Phys. Lett. **B128**, 73 (1983).
[6] C. S. Wood *et al.*, Science **275**, 1759 (1997).
[7] J. Guéna, M. Lintz, and M. A. Bouchiat, Phys. Rev. A **71**, 042108 (2005).
[8] D. Antypas *et al.* Nat. Phys. **15**, 120 (2019)
[9] N. H. Edwards *et al.*, Phys. Rev. Lett. **74**, 2654 (1995).
[10] P. A. Vetter *et al.*, Phys. Rev. Lett. **74**, 2658 (1995).
[11] D. M. Meekhof *et al.*, Phys. Rev. Lett. **71**, 3442 (1993).
[12] M. J. D. MacPherson *et al.*, Phys. Rev. Lett. **67**, 2784 (1991).
[13] E. Gomez *et al.*, Phys. Rev. A **75**, 033418 (2007).
[14] M. N. Portela *et al.*. Hyperfine Interactions **214**, 157 (2013).
[15] B. Batell *et al.*, Phys. Rev. Lett. **107**, 011803 (2011).
[16] G. Feinberg and M. Y. Chen, Phys. Rev. D **10**, 190 (1974).
[17] J. Missimer and L. M. Simons, Phys. Rep. **118**, 179 (1985).
[18] K. Kirch *et al.*, Phys. Rev. Lett. **78**, 4363 (1997).
[19] K. Kirch *et al.*, Phys. Rev. A **59**, 3375 (1999).
[20] F. Wauters and A. Knecht on behalf of the muX collaboration, SciPost Phys. Proc. **5**, 022 (2021).
[21] S. Kanda and K. Ishida, RIKEN Accel. Prog. Rep. **54**, 220 (2021).
[22] K. M. Kojima *et al.*, J. of Phys.: Conf. Ser. **551**, 012063 (2014).
[23] R. Bacher *et al.*, Phys. Rev. A **39**, 1610 (1989).
[24] J. D. Knight *et al*, Phys. Rev. A **27**, 2936 (1983).
[25] D. Abbott *et al.*, PSI Nuclear and Particle Physics Newsletter 1993, p. 61 (1994).

Mu2e DAQ and slow control systems

Antonio Gioiosa[1,2,*], *Richard* Bonventre[3], *Simone* Donati[1,2], *Eric* Flumerfelt[4], *Glenn* Horton-Smith[5], *Luca* Morescalchi[2], *Vivian* O'Dell[4], *Elena* Pedreschi[2], *Gianantonio* Pezzullo[6], *Franco* Spinella[2], *Lorenzo* Uplegger[4], and *Ryan* Rivera[4]

[1]Università di Pisa, I-56127 Pisa, Italy
[2]INFN Sezione di Pisa, I-56127 Pisa, Italy
[3]Lawrence Berkeley National Laboratory, Berkeley, California 94720, USA
[4]Fermi National Accelerator Laboratory, Batavia, Illinois 60510, USA
[5]Department of Physics, Kansas State University, Manhattan, Kansas 66506, USA
[6]Yale University, New Haven, Connecticut, 06520, USA

Abstract. The Mu2e experiment at the Fermilab Muon Campus will search for the coherent neutrinoless conversion of a muon into an electron in the field of an aluminum nucleus with a sensitivity improvement by a factor of 10,000 over existing limits. The Mu2e Trigger and Data Acquisition System (TDAQ) uses *otsdaq* as the online Data Acquisition System (DAQ) solution. Developed at Fermilab, *otsdaq* integrates both the *artdaq* DAQ and the *art* analysis frameworks for event transfer, filtering, and processing. *otsdaq* is an online DAQ software suite with a focus on flexibility and scalability and provides a multi-user, web-based, interface accessible through a web browser. The data stream from the detector subsystems is read by a software filter algorithm that selects events which are combined with the data flux coming from a Cosmic Ray Veto System. The Detector Control System (DCS) has been developed using the Experimental Physics and Industrial Control System (EPICS) open source platform for monitoring, controlling, alarming, and archiving. The DCS System has been integrated into *otsdaq*. A prototype of the TDAQ and the DCS systems has been built at Fermilab's Feynman Computing Center. In this paper, we report on the progress of the integration of this prototype in the online *otsdaq* software.

1 Introduction

Lepton Flavor Violation (LFV) has been observed in the neutral sector in neutrino oscillations but not in the charged sector. In the Standard Model (SM), the predicted branching fractions of Charged Lepton Flavor Violating (CLFV) processes are below 10^{-50}. The observation of CLFV would thus provide unambiguous evidence for the existence of New Physics beyond the SM. The Mu2e experiment at Fermilab will search for the coherent neutrinoless muon-to-electron-conversion in the field of an aluminum nucleus ($\mu^- + {}^{27}_{13}\mathrm{Al} \rightarrow e^- + {}^{27}_{13}\mathrm{Al}$) [1]. The expected Mu2e single event sensitivity (SES) is:

$$R_{\mu e} \equiv \frac{\Gamma(\mu^- N(A,Z) \rightarrow e^- N(A,Z))}{\Gamma(\mu^- N(A,Z) \rightarrow \nu_\mu N(A,Z-1)^*)} = 3 \times 10^{-17}. \tag{1}$$

*e-mail: antonio.gioiosa@df.unipi.it

The current world's best limit $R_{\mu e} < 7 \times 10^{-13}$ (on gold) is from the SINDRUM II experiment at Paul Scherrer Institut [2]. In addition to Mu2e, the COMET experiment in preparation at J-PARC has an expected SES of 3×10^{-15} for Phase-I and $O(10^{-17})$ for Phase-II (on aluminum) [3], while the DeeMe experiment, also in preparation at J-PARC, has an expected SES of 10^{-13} (on carbon) [4].

The Mu2e apparatus includes three superconducting solenoids:

1. The Production Solenoid, where the 8 GeV pulsed proton beam (period of 1.7 μs) hits the tungsten production target and produces mostly pions;

2. The Transport Solenoid, which serves as a decay "tunnel" for the pions and performs a charge and momentum selection, thus producing the low-momentum μ^- beam;

3. The Detector Solenoid, where muons are stopped in the aluminum stopping target, form muonic atoms, then decay to 105 MeV/c^2 electrons, which are detected by state-of-the-art detectors for tracking and energy reconstruction.

The Mu2e Trigger and Data Acquisition System (TDAQ) is a streaming system with a software-only trigger designed to satisfy the following requirements [5][6]:

- Provide efficiency better than 90% for the conversion electron signal;
- Keep the total trigger rate below a few kHz - equivalent to approximately 7 PB/year of total data rate (to keep the collected data less than 21 PB in 3 years of running);
- Keep the processing time below 5 ms/event (that is the spill off time between two spills where the beam will be on).

To achieve these goals and allow for a higher off-detector data rate (desired with a 'software-only' trigger architecture), the Mu2e Data Acquisition System (DAQ) is based on a streaming readout. This means that all detector data are digitized, zero-suppressed in the front-end electronics and then transmitted off the detector to the DAQ. In this paper, we present the Mu2e Trigger and Data Acquisition System (TDAQ) and the Detector Control System (DCS) prototypes built at Fermilab's Feynman Computing Center. We also present the Mu2e online DAQ software suite *otsdaq* designed and developed at Fermilab. We report at end the integration of the DCS system into the online *otsdaq* software.

2 The TDAQ System

The Mu2e Trigger and Data Acquisition System (TDAQ) provides the necessary infrastructure to collect digitized data from the tracker, calorimeter, cosmic ray veto and monitor the beam status. The TDAQ employs 36 dual-CPU servers to handle a total rate of 192,000 proton pulses per second and an average of 5,400 events per second per server. According to preliminary estimates and DAQ bandwidth requirements, the detectors generate approximately 120 kB of zero-suppressed data per proton pulse for a resulting average total data rate of about 20 GB/s when beam is present [5]. Figure 1 shows the global TDAQ architecture.

The Run Control Host receives beam status and timing information from the Accelerator Control network. The Control Fanout (CFO) module in the Run Control Host is responsible for generating and synchronizing Readout Requests.

As its main function, each Read Out Controller (ROC), that has a large front end buffer to average over off-spill time, continuously streams out the zero-suppressed data collected between two proton pulses from the detectors to the DTCs (Data Transfer Controller). The data of a given time-frame are then collected in a single server using a 10 Gb/s switch. Then, the online reconstruction starts and a trigger decision is made. If an event gets triggered, the

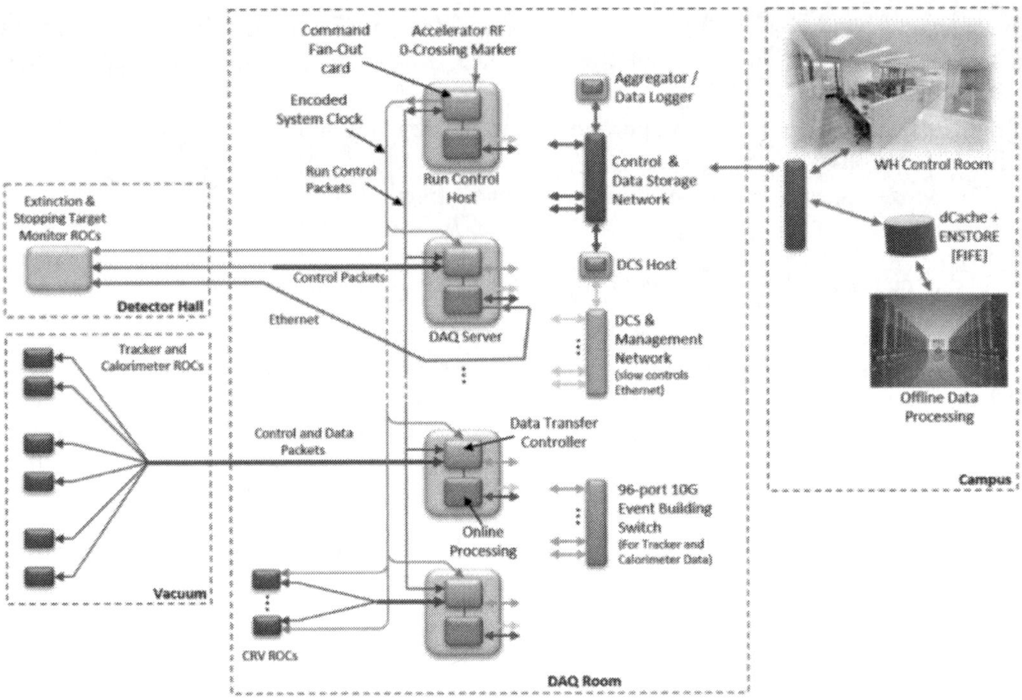

Figure 1. Mu2e TDAQ architecture and components diagram.

data from the cosmic ray veto (CRV) are pulled and aggregated into a single data stream. The DAQ servers filter these events (aggregator/data logger) and forward a small subset of them to the offline storage. A total of 497 ROCs and 83 DTCs will be used.

Figure 2 shows a scheme of the Mu2e data readout topology. In the Data Concentrator layer, ROCs data are directly collected by the DTCs via optical links. Using the 10 Gb/s ethernet network switch, events are then formed and triggered in the Event Builder layer where DTCs are involved in a round-robin process. At the end, events are moved via a Peripheral Component Interconnect Express (PCIe) bus to the Storage Layer for further filtering and data logging.

The TDAQ employes *otsdaq* as a software solution. Developed at Fermilab, it uses the *artdaq* [7, 8] and *art* [9] software as event filtering (data transfer, event building and event reconstruction) and processing frameworks. *otsdaq* includes a run control system using the data acquisition software XDAQ [10] implemented for the development and calibration-mode runs at CMS. The *otsdaq* development for Mu2e follows two main directions: server side and web side. The server side is developed in C++. The specific code for Mu2e is added through plugins (C++ classes inheriting from the appropriate base class). The web side (directly accessible to the end user through a web browser) is developed in HTML and JavaScript. Custom code for Mu2e is added in the form of web-apps through html files.

The Mu2e physics triggers identify signal event candidates [6]. It is implemented as a series of software filters applied after each step of track reconstruction. The total trigger rate is expected not to exceed 700 Hz [6]. With the *artdaq* framework, it is possible to limit the offline data storage to less than 7 PB/year with a reduction factor of about 100 at the event building level [11].

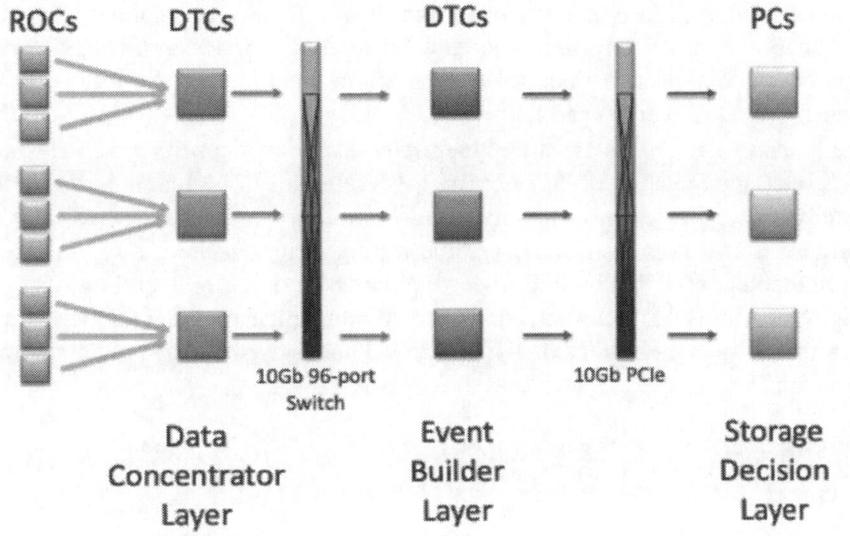

Figure 2. Mu2e data readout topology.

3 The Detector Control System

The Detector Control System (DCS) function is to monitor the detectors status and operational conditions. For each subsystem, the DCS allows to display real-time event and detector data via Graphical User Interfaces (GUIs) and archive the monitoring data to disk.

Mu2e selected EPICS (Experimental Physics and Industrial Control System) for the DCS slow control and monitoring software[12]. EPICS, with the Control System Studio (CSS) GUI software, is an open source framework originally developed at Argonne and Fermilab and now used in numerous experiments [13]. An Input Output Controller (IOC), running for each subsystem on a central DAQ server, will provide channels for all data [14]. The total number of slow control quantities is expected to be of the order of thirty thousand. On average, these quantities will be updated approximately twice per minute, for a resulting generated data rate of 10 kB/s.

As part of the DCS, *otsdaq* delivers slow control data from the DTCs and ROCs to EPICS. The *otsdaq* allows the user to monitor and interact with the DAQ hardware and the other devices managed by EPICS to:

1. Observe Process Variables (PVs) such as settings, alarms, warnings, readouts, timestamps, status;
2. Interact through a web interface that is lightweight, user-friendly, ready to use, customizable;
3. Implement custom handling of PV alarms integrated with the TDAQ state machine transitions.

The design of *otsdaq* involved C++ and web-app applications to include the Mu2e slow control monitoring, alarm handling, and TDAQ hardware and online daq slow control entities writing in EPICS. From *otsdaq* it is possible to handle alarm propagation from EPICS IOC to automated plug-in threads or to the web user interface and, ultimately, the users.

To connect *otsdaq* to EPICS, a C++ interface has been developed and it uses the EPICS Channel Access Client Library functions and Postgres database connections to read/write data. This interface allows for monitoring and alarm handling for the following PV data: Value, Alarm (Status, Severity) and Settings.

Figure 3 shows the C++ classes and hierarchy design of the *otsdaq* EPICS interfacing. The EPICS Interface is a child of the ots::ControlsVInterface Class that inherits Configurable and VStateMachine Classes (not shown). It allows to handle the State Machine transitions halt, configure, start, stop, pause and resume during State Machine traversal. Moreover, it is "Configurable", so it is possible to set PV names and alarms handling, using the *otsdaq* configuration tools. All these classes stem from the underlying XDAQ (used in *otsdaq*) framework that defines the actions on the controlled nodes by such state machine transitions.

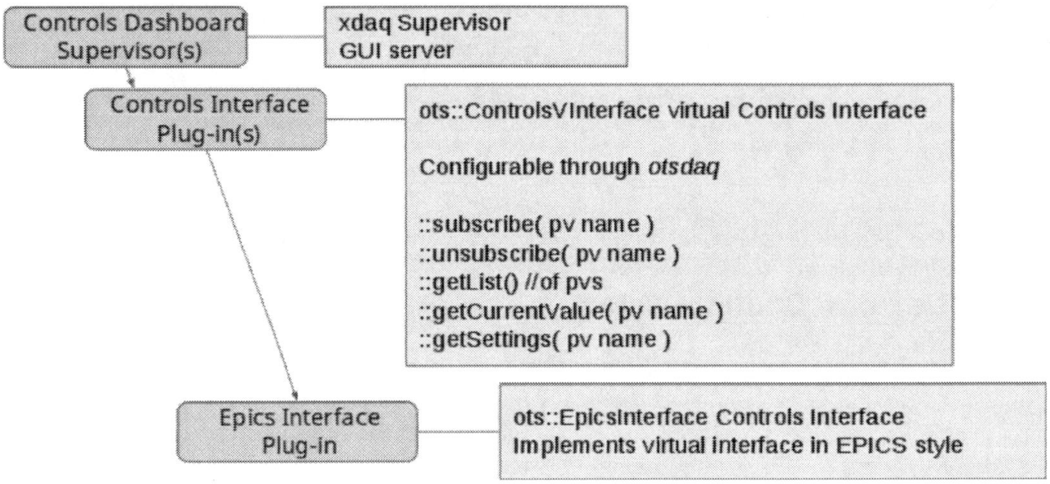

Figure 3. *otsdaq* C++ classes designed for EPICS interfacing. ots::ControlsVInterface Class inherits Configurable and VStateMachine Classes (not shown).

otsdaq uses an *artdaq* interface to EPICS CA in order to write EPICS PVs for DAQ hardware and metrics.

Figure 4 shows the *otsdaq* slow control web GUI C++ classes diagram and web-app connections that explicit the GUI design and functioning. The Supervisor C++ Class ControlsDashboardSupervisor manages the *otsdaq* web clients requests, dispatching backward the EPICS Interface instance data collections. The HTML object SlowControlsDashboard forms the dashboard window directly handled by the end user which may have tools such as widgets editor and file pages access to retrieve/save slow control pages from/to the web server.
We developed the dashboard GUI to read/write slow control pages in the XML format compatible with CSS Phoebus version of EPICS GUI software.

4 Conclusions

In this article, we have presented the Trigger and Data Acquisition System (TDAQ) and Detector Control System (DCS) currently being developed for the Mu2e experiment at Fermilab.

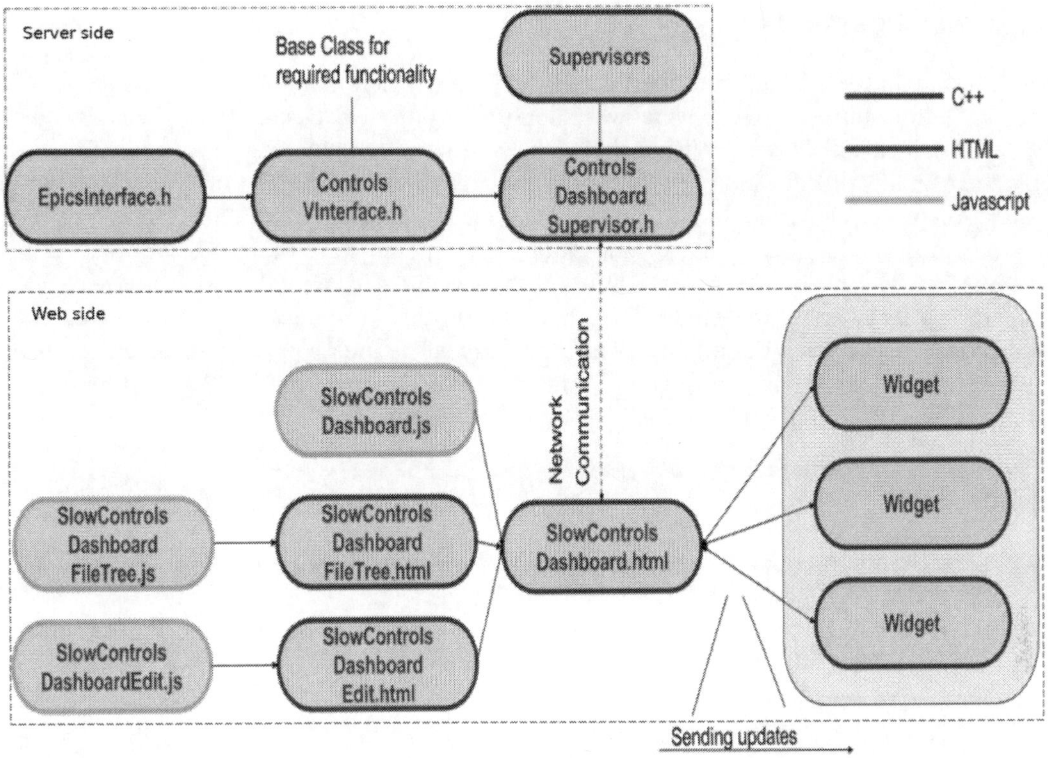

Figure 4. *otsdaq* slow control web GUI C++ classes diagram and web-app connections scheme. General information on Interface plugin classes used in *otsdaq* is reported in section 2.

The TDAQ system uses the online DAQ software suite *otsdaq* developed at Fermilab to provide a high level of flexibility and scalability. We have reported on the preliminary results of the system performance. The Detector Control System (DCS) system uses the open source framework EPICS developed at Argonne and Fermilab and widely employed in a number of experiments, including CMS. The *otsdaq* system includes a part of DCS that communicates with EPICS. A run control GUI has been developed and integrated in *otsdaq* to provide a multi-user, web-based control and monitoring dashboard.

A test-stand equipped with the *otsdaq* web-based GUI was set up at Fermilab to test TDAQ and its timing performance. First level results of tests show that the requirement to achieve a total processing time/event below 5 ms is matched only if one event-builder/DAQ server is implemented [6]. With two or more event-builders/DAQ servers implemented, the competition for the shared resources and the increased architecture complexity degrades the system performance. The system is still being developed and a number of possible alternative solutions to mitigate these effects are currently being explored.

Concerning data reduction performances, the TDAQ system is able to limit the offline data storage to less than 7 PB/year. With the *artdaq* framework, it is possible to satisfy the requirement with a reduction factor of about 100 at the event building level [11]. Moreover, by default, Mu2e data are saved in *art* ROOT format to further reduce the amount of pre-processing necessary for offline analysis.

Acknowledgements

We are grateful for the vital contributions of the Fermilab staff and the technical staff of the participating institutions. This work was supported by the US Department of Energy; the Istituto Nazionale di Fisica Nucleare, Italy; the Science and Technology Facilities Council, UK; the Ministry of Education and Science, Russian Federation; the National Science Foundation, USA; the Thousand Talents Plan, China; the Helmholtz Association, Germany; and the EU Horizon 2020 Research and Innovation Program under the Marie Sklodowska-Curie Grant Agreement No. 690835, 734303, 822185, 858199, 101003460. This document was prepared by members of the Mu2e Collaboration using the resources of the Fermi National Accelerator Laboratory (Fermilab), a U.S. Department of Energy, Office of Science, HEP User Facility. Fermilab is managed by Fermi Research Alliance, LLC (FRA), acting under Contract No. DE-AC02-07CH11359.

References

[1] R.H. Bernstein *et al*, "Charged Lepton Flavor Violation: An Experimenter's Guide," Phys. Rep., 532, 2, 27-64, 2013.

[2] W.H. Bertl *et al*, "A Search for muon to electron conversion in muonic gold," Eur. Phys. J., C47 p. 337-346, 2006. doi:10.1140/epjc/s2006-02582-x.

[3] M. Lee, "Comet Muon conversion experiment in J-PARC," Frontiers in Physics, v. 6 p. 133, 2018. doi:10.3389/fphy.2018.00133.

[4] N. Teshima, "Status of the DeeMe Experiment, an Experimental Search for μ-e Conversion at J-PARC MLF," PoS, v. NuFact2019 p. 082, 2020. doi:10.22323/1.369.0082.

[5] L. Bartoszek *et al*, "Mu2e Technical Design Report", 1501.05241, arXiv, physics.ins-det, 2015.

[6] D. Brown, G. Pezzullo, P. Murat, "The Track finder algorithm for the Trigger system of the Mu2e experiment at Fermilab," Connecting The Dots, Princeton, 2020. *https://doi.org/10.5281/zenodo.4088480*

[7] K. Biery, E. Flumerfelt, J. Freeman, W. Ketchum, G. Lukhanin, A. Lyon, R. Rechenmacher, R. Rivera, L. Uplegger and M. Votava, "Flexible and Scalable Data-Acquisition Using the artdaq Toolkit," [arXiv:1806.07250 [physics.ins-det]], 2018.

[8] K. Biery, C. Green, J. Kowalkowski, M. Paterno and R. Rechenmacher, "artdaq: An event filtering framework for Fermilab experiments," doi:10.1109/RTC.2012.6418358, 2012.

[9] C. Green, J. Kowalkowski, M. Paterno, M. Fischler, L. Garren and Q. Lu, "The Art Framework," J. Phys. Conf. Ser. 396, 022020 (2012) doi:10.1088/1742-6596/396/2/022020.

[10] G. Bauer et al., "The CMS data acquisition system software," J. Phys. Conf. Ser., vol. 219, p. 022011, 2010.

[11] Kurt Biery et al 2017 J. Phys.: Conf. Ser. 898 032013.

[12] https://epics-controls.org/, 2020.

[13] https://epics-controls.org/epics-users/projects/, 2020.

[14] "EPICS Record Reference Manual" http://www.ing.iac.es/ docs/external/epics/recordref.pdf.

Proposal for new measurements of muonic helium hyperfine structure at J-PARC

Seiso Fukumura[1], *Patrick* Strasser[2], *Takashi* Ino[2], *Takayuki* Oku[3], *Takuya* Okudaira[1], *Sohtaro* Kanda[2], *Masaaki* Kitaguchi[1], *Koichiro* Shimomura[2], *Hirohiko M.* Shimizu[1], *Hiroyuki A.* Torii[4], and *Shoichiro* Nishimura[2]

[1]Department of Physics, Nagoya University, Furo-cho, Chikusa-ku, Nagoya, 464-8601, Japan
[2]High Energy Accelerator Research Organization (KEK), 1-1 Oho, Tsukuba, Ibaraki 305-0801, Japan
[3]Advanced Science Research Center, Japan Atomic Energy Agency (JAEA), 2-4 Shirane Shirakata, Tokai-mura, Naka-gun, Ibaraki 319-1195, Japan
[4]School of Science, The University of Tokyo, 7-3-1 Hongo, Bunkyo, Tokyo 113-0033, Japan

Abstract. The measurement of the ground state hyperfine structure of muonic helium has the potential to improve the precision of the mass of the negative muon by a factor of 50 or more. The mass of the negative muon is very important because it enables us to test the CPT theorem by comparison with positive muon mass. We aim to measure the hyperfine structure of muonic helium with a precision 1000 times higher than previous experiments [1,2] using the high-intensity muon beam at J-PARC and have already obtained results better than the current precision in zero-field measurements in a test experiment in March 2021. To further improve the precision, we plan to measure in a high magnetic field and incorporate a technique that can produce highly polarized muonic helium atom [3]. In this paper, we will report on these developments.

1 Introduction

The muonic helium atom (^{4}Heμ^-e^-) is an atom in which one of the two electrons of the ^{4}He atom is replaced by a negative muon. Since the orbital radius of the negative muon is about 400 times smaller than the Bohr radius, we can treat (^{4}Heμ^-)$^+$ as a pseudo-nucleus that has a magnetic moment similar to that of a negative muon (μ_{μ^-}), and consider ^{4}Heμ^-e^- as a hydrogen-like atom. The ground state hyperfine structure (HFS) of muonic helium is caused by the interaction between the magnetic moments of the electron and the negative muon. The HFS of muonic helium is similar to muonium, which is a purely leptonic system consisting of positive muon and electron, and it can be measured using the same apparatus and technique as for muonium ground state HFS [4-5].

Since muonic helium is a three-body system, precise HFS measurements of muonic helium will be a powerful probe to test and improve the theory of quantum three-body systems [6]. Measurements at high magnetic field will also allow us to precisely determine the magnetic moment and mass of negative muon, and to test the CPT theorem by comparison with the positive muon. The ground state HFS of muonic helium has been measured with a precision of 13 ppm at zero field [1] and 6.5 ppm at high field [2]. The precision of the magnetic moment ratio of negative muons to protons has been determined as 47 ppm [2]. These uncertainties are mainly statistical uncertainty. The reason for the large statistical uncertainty is not only the beam intensity but also the loss of muon spin polarization (~100% → ~ 6%) due to Auger transitions and Stark mixing associated with the formation of muonic helium [7-12]. For theoretical values, the higher-order QED effect has not been taken into account so far, and uncertainty of about 130 ppm has been estimated [13-14].

For the mass of negative muons (m_{μ^-}), the highest precision is currently 3.1 ppm obtained by precise measurement of characteristic X-rays of muonic Si [15]. In contrast, the mass of the positive muon is determined to be 120 ppb from the measurement of the muonium ground state HFS [16]. In other words, the verification of the CPT theorem for muons is presently limited to the level of 3 ppm. In addition, the magnetic moment of the negative muon (μ_{μ^-}) is necessary for the determination of the magnetic moment anomaly of negative muons at Brookhaven National Laboratory, but due to the large uncertainty of the μ_{μ^-} measurement, it is currently calculated using the magnetic moment of positive muons [17]. Furthermore, muonium 1S-2S transition frequency measurements by laser spectroscopy aiming to determine the positive muon mass in O(1 ppb) are planned by the MuMASS collaboration at PSI [18] and an experiment at J-PARC [19-20].

Currently, we are planning to measure the ground state HFS of muonic helium at zero field and high field of 1.15 T in O(10 ppb) and m_{μ^-} in O(100 ppb). This will be achieved by using the high-intensity negative muon beam from the H-Line of the Muon Science Establishment (MUSE) in the Materials and Life Science Experimental Facility (MLF) at the Japan Proton Accelerator Research Complex (J-PARC), which will be operated from 2022, and repolarizing the muon spin using the Spin Exchange Optical Pumping (SEOP) technique. In addition to the improvement of the measurement precision, we are also investigating the improvement of the theoretical values in collaboration with theorists at RIKEN and Tohoku University.

2 Experimental method and latest result

As shown in Fig. 1, the ground state of the muonic helium atom splits into four sub-levels under a static magnetic field due to the Zeeman effect. By applying a microwave field and measuring the transition frequencies between the levels, the ground state HFS $\Delta\nu$ of the muonic helium atom can be obtained directly at zero field and indirectly as the sum of ν_{12} and ν_{34} at high field. The difference between ν_{12} and ν_{34} can also be used to determine the magnetic moment and mass of the negative muon. Fig. 2 shows the conceptual diagram of the measurement system. The measurement is mainly performed in three steps: (1) formation of muonic helium atoms, (2) state transition induced by microwave field, (3) measurement of the asymmetry of the decay electron emission.

2.1 Formation of muonic helium atoms

A polarized negative muon beam (the polarization is about 30–50% for cloud muons in the H-Line, and about 90% for backward decay muons in the D-Line) is injected into a microwave cavity installed in a the gas chamber. The gas chamber is filled with ^{4}He gas as target and a small amount (~2%) of CH_4 as an electron donor. The muon stops in the ^{4}He gas, is captured by a ^{4}He nucleus, and forms a $(^4\mathrm{He}\mu^-)^+$ ion in the ground state through Auger and radiative transitions. The $(^4\mathrm{He}\mu^-)^+$ ion forms a muonic helium atom $(^4\mathrm{He}\mu^-e^-)$ through charge exchange with the surrounding CH_4. During the muon capture process, the polarization of the muon decreases to about 6% of the incident value due to Auger transitions and Stark mixing.

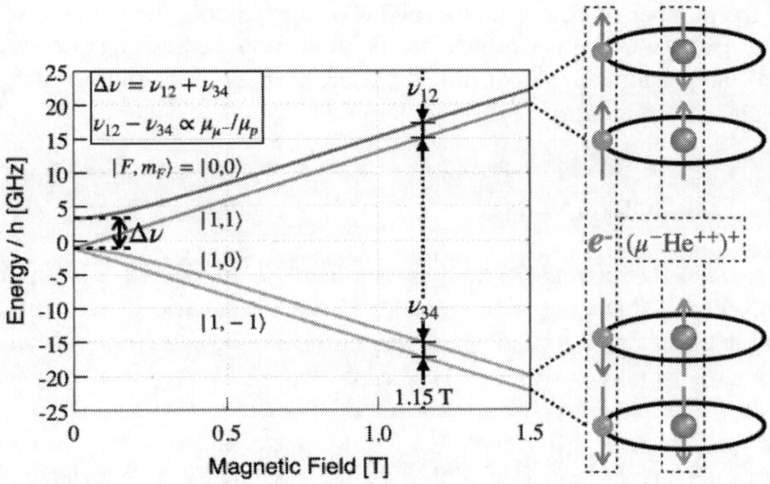

Figure 1. The energy level diagram for muonic helium. F is the total angular momentum, and m_F is the quantization axis component of F.

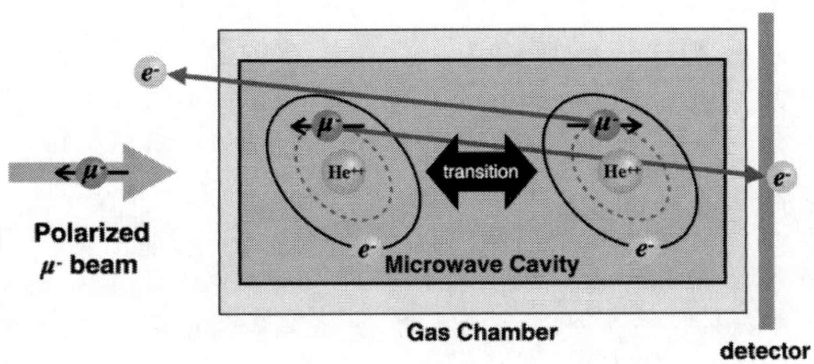

Figure 2. Schematic view of the experimental setup. This apparatus is either enclosed in a magnetic shield box for zero-field measurements or inserted in a large superconducting solenoid magnet for high-field measurements.

2.2 State transition induced by microwave field

When a microwave field is applied to $^4\mathrm{He}\mu^- e^-$, the state transitions occur with a probability corresponding to the difference between the transition frequency and the microwave field frequency, flipping the spin of the negative muon in muonic helium.

2.3 Measurement of the asymmetry of the decay electron emission

The decay electrons tend to be emitted antiparallel to the muon spin. Using backward decay muons (spin antiparallel to the muon momentum direction), when the applied microwave frequency is close to state transition frequency, the amount of decay electrons emitted in the downstream direction is reduced because of the spin-flip. The electrons are detected by a segmented scintillation detector downstream of the gas chamber. The resonance curve is obtained by sweeping the frequency of the applied microwave field and measuring the difference in the number of detected electrons with and without the microwave field. The center of the resonance curve corresponds to the transition frequency between states. The ground state HFS of muonic helium is directly determined by measuring $\Delta\nu$ at zero field and indirectly obtained by summing ν_{12} and ν_{34} at high field.

2.4 A result of preliminary tests

The ground state HFS of muonic helium at zero field was measured using a pulsed muon beam at J-PARC MUSE D-Line in March 2021. The ^{4}He gas was mixed with 2% of CH_4 and the total pressure was 4 atm. The momentum of the negative muon beam was set to 27 MeV/c to maximize the amount of ^{4}Heμ^-e^- produced. We are still analyzing the measurement results. The asymmetry spectra of the decay electrons with an applied microwave field frequency of 4465.050 MHz and 4465.650 MHz are shown in Fig. 3. When the applied microwave frequency was 4465.050 MHz, the asymmetry was negative. In contrast, when the applied microwave frequency was 4465.650 MHz, the asymmetry was nearly zero. This is because the ground state HFS of muonic helium is near 4465.050 MHz. Considering the beam intensity, the precision from this measurement is expected to be comparable to the current highest precision of 6.5 ppm at high magnetic field [2].

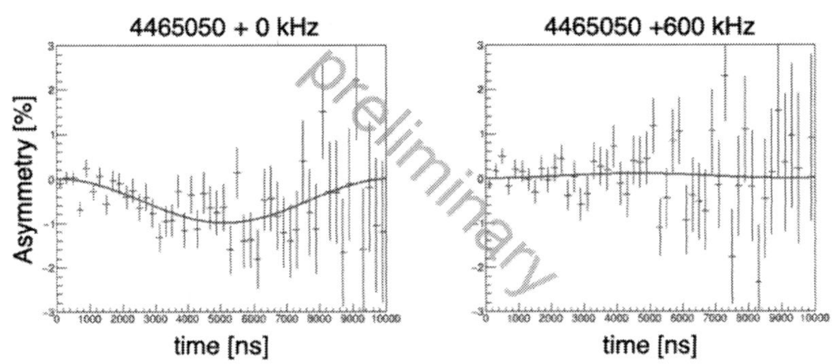

Figure 3. Asymmetries of electron emitted direction applied microwave frequency detuned 0 kHz and 600 kHz from 4465.050 MHz. The asymmetry is defined as the $N_{on}/N_{off} - 1$. N_{on} is the number of detected electron with microwave field, and N_{off} without, respectively.

3 Toward higher precision

3.1 H-Line

To obtain more statistics, we will use the J-PARC MUSE H-Line, which is expected to start operation in 2022. The beam intensity of the H-Line is 10 times higher than that of the D-Line. H-Line will enable us to perform measurements under high magnetic field using superconducting magnets. At high magnetic field, decay electrons are focused along the longitudinal direction, so more electrons can be detected with the same detector as in zero-field measurements. After 100 days of measurement in the H-Line under a magnetic field of 1.15 T, the measurement precision is expected to improve by a factor of ten from the previous measurement [2] to 600 ppb.

3.2 Spin Exchange Optical Pumping for Muonic Helium repolarization

The statistical uncertainty in $\Delta\nu$ is presently limited by the negative muon depolarization associated with the formation of $^4\text{He}\mu^-e^-$. We aim to further improve the measurement precision by repolarizing the negative muons through SEOP of $^4\text{He}\mu^-e^-$. Before $^4\text{He}\mu^-e^-$ repolarization, the outermost shell electrons of alkali metal atoms are polarized by optical pumping. Then a muonic helium atom collides and spin-exchanges with an electron-polarized alkali metal atom and the electron of $^4\text{He}\mu^-e^-$ is polarized. Afterward, the electron polarization is shared with the pseudo-nucleus via the hyperfine interaction, and polarizing μ^-. In a previous experiment [3], the spin exchange between Rb and $^4\text{He}\mu^-e^-$ was performed and the μ^- was repolarized in $^4\text{He}\mu^-e^-$ from 6% to 44% . The spin-exchange cross section was obtained as $\sigma_{\text{SE}} = (4.47 \pm 0.67 \pm 0.82) \times 10^{-14}$ cm^2. Using this value in a Monte Carlo simulation, the time evolution of the polarization of μ^- at 1 mT and 200°C is expected to be as shown in Fig.4. In this simulation, we assume the number of Rb is 4.37×10^{14} cm^{-3} and the spins of electron and muon are determined according to the ratio of the square of Clebsch-Gordan coefficients. This means that if we measure decay electrons 2000 ns after each muon beam pulse is injected, we can measure the ground state HFS of muonic helium with almost 100% polarized μ^-. In practice, the achievable polarization decreases with laser power and gas impurities. Assuming a 7-fold improvement in the polarizability of μ^- based on the previous study [4], we can expect a direct improvement by a factor of seven in measurement precision.

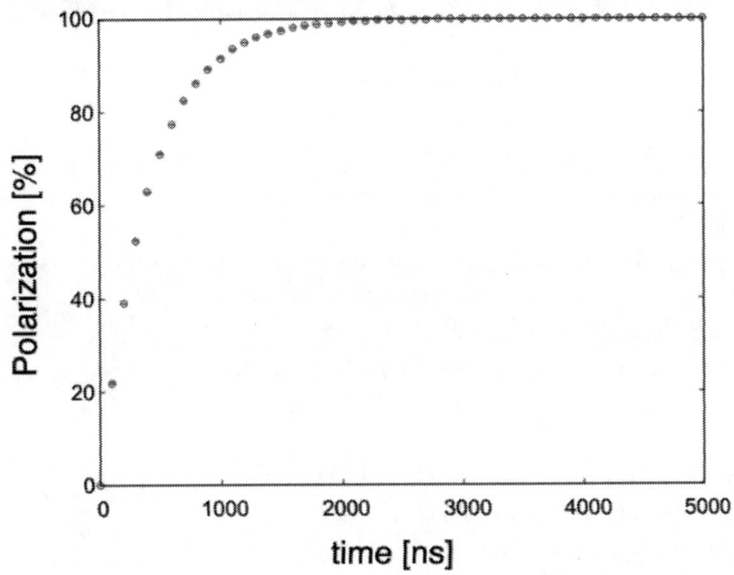

Figure 4. Simulation result of the pseudo-nucleus polarization of muonic helium with electron-polarized Rb at 1 mT and 200°C .

4 Summary

Precise measurements of muonic helium HFS are important for the verification of quantum three-body theory and CPT theorem. We are currently working on zero-field measurements at the MUSE D-Line, and preparations for high-field measurements at the H-Line are underway.

We are also planning to repolarize the muonic helium to further improve the measurement precision by using SEOP.

5 Acknowledgements

The muon experiment at the Materials and Life Science Experimental Facility of the J-PARC was performed under a user program (2020B0333). This work was supported by the JSPS KAKENHI (Grant-in-Aid for Scientific Research (A) 21H04481).

References

[1] H. Orth, et al., Phys. Rev. Lett. **45**, 1483 (1980)
[2] C. J. Gardner, et al., Phys. Rev. Lett. **48**, 1168 (1982)
[3] A. S. Barton, et al., Phys. Rev. Lett. **70**, 758 (1993)
[4] S. Kanda, et al., Phys. Lett. B **815**, 136154 (2021)
[5] P. Strasser et al., JPS Conf. Proc. **21**, 011045 (2018).
[6] E. W. Schmid and H. Ziegelmann, *The Quantum Mechanical Three-Body Problem*, ed. H. Stumpf (Pergamon Press, 1974)
[7] R. A. Mann and M. E. Bose, Phys. Rev. **121**, 293 (1961)
[8] P. A. Souder et al., Phys. Rev. A **22**, 33 (1980).
[9] I. M. Shmushkevich, Nucl. Phys. **11**, 419 (1959)
[10] N. C. Mukhopadhyay, Phys. Rep. **30C**, 1 (1977).
[11] T. Koike and Y. Akaishi, Nucl. Phys. A **639**, 521c (1998)
[12] V. W. Hughes and G. zu Putlitz, in *Quantum Electro-dynamics*, ed. T. Kinoshita (World Scientific, Singapore, 1990) p. 822.
[13] K. Pachucki, Phys. Rev. A **63**, 032508 (2001)
[14] A. A. Krutov and A. P. Martynenko, Phys. Rev. A **78**, 032513 (2008)
[15] I. Beltrami, et al., Nucl. Phys. A **451**, 679 (1986)
[16] W. Liu, et al., Phys. Rev. Lett. **81**, 711 (1999)
[17] G. W. Bennett, et al., Phys. Rev. A **92**, 161802 (2004)
[18] P. Crivelli, Hyperfine Interact **239**, 49, (2018)
[19] C. Zhang, et al., JPS Conf. Proc. **33**, 011125 (2021)
[20] S. Uetake, et al., High Energy News, Vol. 39, No. 4, 170-181 (2021) (in Japanese)

EPJ Web of Conferences **262**, 01013 (2022)
EXA2021
https://doi.org/10.1051/epjconf/202226201013

Upgrade of the scintillating bars detector for the ASACUSA experiment

Giovanni Costantini[1,2], *Luca* Giorleo[3], *Giulia* Gosta[1,2,*], *Marco* Leali[1,2], *Valerio* Mascagna[1,2], *Stefano* Migliorati[1,2], *Michela* Prest[4,5], *Federico* Ronchetti[4,5], *Luigi* Solazzi[3,2], *Erik* Vallazza[5], and *Luca* Venturelli[1,2]

[1]Dipartimento di Ingegneria dell'Informazione (DII), Università degli Studi di Brescia, Brescia, Italy
[2]INFN - Sezione di Pavia, Pavia, Italy
[3]Dipartimento di Ingegneria Meccanica e Industriale (DIMI), Università degli Studi di Brescia, Brescia, Italy
[4]Dipartimento di Scienze e Alta Tecnologia (DiSAT), Università degli Studi dell'Insubria, Como, Italy
[5]INFN - Sezione di Milano-Bicocca, Milano, Italy

Abstract. The upgrade of the scintillating bars detector, used in the experiments of the ASACUSA Collaboration at the CERN Antiproton Decelerator is presented. The detector consists of several modules, each one made of ≈1m long scintillating bars, which allow the detection of the charged particles produced in the annihilations of antiprotons and antihydrogens. The mechanics, the electronic readout and the data acquisition system upgrade are described.

1 Introduction

The ASACUSA collaboration (Atomic Spectroscopy And Collisions Using Slow Antiprotons) performs experiments with low energy antiprotons at the CERN Antiproton Decelerator (AD). The detector we are presenting has been used by the collaboration in different ways: to measure the antiproton annihilation cross sections on different target nuclei and energies [1–5], or by recording the annihilation rate and the vertex position of antiprotons ($\bar{p}$s) and antihydrogens ($\bar{H}$s) in the antihydrogen experimentation, where it resulted to be fundamental to demonstrate the first $\bar{H}$ production in a cusp trap [6]. The detector in the upgraded configuration will be used, in particular, in the experiment regarding the measurement of the hyperfine splitting of the ground-state (HFS-GS) of antihydrogen with the aim of performing a test of the fundamental CPT symmetry between matter and antimatter [7]. In figure 1 the experimental setup for the $\bar{H}$ production and measurement is shown.

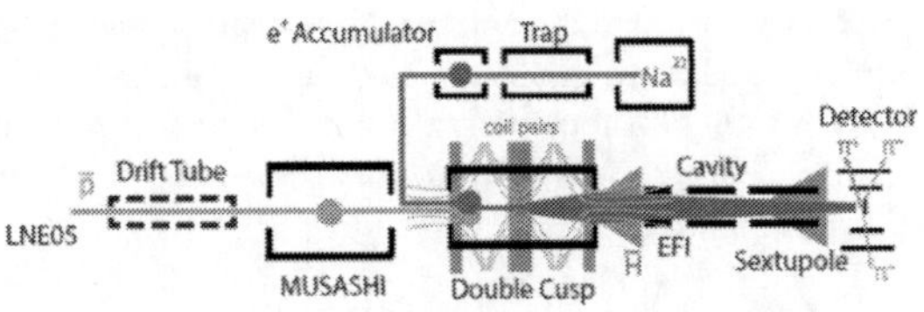

Figure 1: ASACUSA experimental setup for antihydrogen measurement.

The MUSASHI trap (see figure 1) cools and stores 100 keV antiprotons delivered by the AD with the ELENA ring (Extra Low ENergy Antiproton) [8]. Positrons, instead, are obtained from a ^{22}Na source and cooled and stored in the positron accumulator. The double-Cusp trap [9] is made by a multi-ringed electrode trap [10] housed within a magnetic field produced by a pair of superconducting coils in an anti-Helmholtz configuration [11]. Here, positrons and antiprotons mix, and the cusp magnetic field focuses, along the apparatus axis, the low-field seeking (LFS) cold ground state antihydrogen atoms. The polarised $\bar{H}$ atoms escape the trap and enter the spectrometer consisting of a microwave cavity [12, 13] to induce hyperfine transitions, followed by a state-analysing sextupole magnet that focuses the low-field seeking states and defocusses the high-field seekers. A detector records the $\bar{H}$ annihilation signal at the end of the beam-line as a function of the microwave frequency applied in the cavity [14–16]. The challenge lies in producing an intense, focused and polarised source of $\bar{H}$ atoms in the ground state. The 3D track detector, described in this work, is used to monitor the annihilations of antiproton and antihydrogen atoms.

2 Detector of antiproton annihilation

The detector consists of 8 planes made of a different number of scintillating bars (from 30 to 60), which were provided by FNAL [17]. The length and section of each bar are 96 cm and 1.5×1.9 cm^2 respectively. They are extruded plastic scintillators composed of Polystyrene Dow Styron 663 W + 1 % PPO + 0.03% POPOP and white TiO2 coating [18]. A hole of 2.5 mm diameter is grooved along their length to host a 1 mm diameter Y-11 type WLS fiber by Kuraray[1] glued with E30 epoxy. The emission peak of the scintillator and the absorption one of the WLS

*e-mail: giulia.gosta@unibs.it

[1]https://www.kuraray.com/

fibers, 420 nm and 430 nm respectively, are well-matched and the light output of such a system has been characterized in Ref. [18].

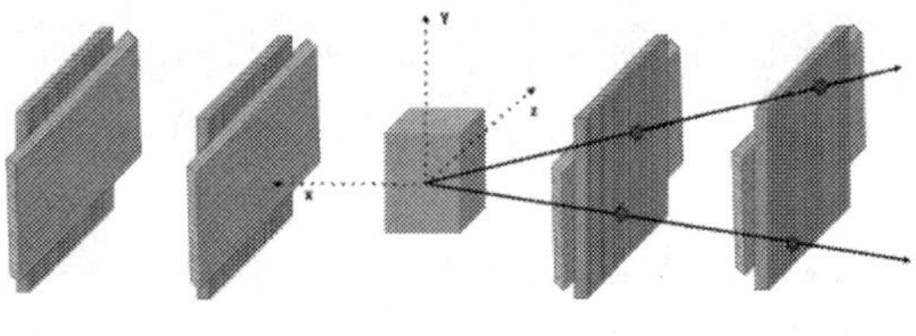

Figure 2: In a typical configuration, the detector consists of 8 planes made of a different number of scintillating bars (from 30 to 60). The planes are coupled so that the two series of bars result orthogonal and they are located around the cusp-trap in order to determine the vertex position in the 3D space.

The planes detect the charged particles crossing the scintillating bars (mostly pions emerging from antiprotons annihilations) and they are a useful tool for tuning the trap parameters during the experiment. The planes are coupled so that the two series of bars result orthogonal to give a bi-dimensional (XY) information of the particle position. The XY-modules are located around the cusp-trap as illustrated in figure 2. In this configuration, the vertex position of $\bar{p}$s and $\bar{H}$s annihilation can be determined in the 3D space.

In the past detector configuration, the light from the WLS fibers of a single plane was collected by 2 Hamamatsu[2] 64 channel H7546-B multianode PhotoMultipliers (PMTs) where only half of the available channels were used in order to reduce the cross-talk between contiguous channels. The fibers were faced to the PMTs by using aluminum plates with holes spaced in a checkboard array. The electronics readout was performed by dedicated front-end boards (FEB) equipped with a MAROC3 ASIC. The FEBs host 2 FPGAs (Altera Cyclone II), a 12 bit ADC and present a dedicated socket to connect one single PMT. The analog signals from the PMT are amplified by a pre-amplifier with a variable gain (8 bit). The signals are then fed to two different chains: the first one (analog output) consists of a slow shaper and a sample & hold circuit, the second one (digital output) is formed by a fast shaper and a discriminator.
Due to the difficulty to keep a good coupling between fibers and PMT in front of many detector movements, a loss of signal and a non-uniform efficiency were observed. Moreover, the fact that the fibers are collected in the same point and read by a single PMT caused non-natural folding of the fibers and as a consequence, the possibility of damaging the fibers. For these reasons, an upgrade was planned.

[2]https://www.hamamatsu.com/eu/en/index.html

3 DANTE: Detector for ANnihilation Tracking Experiment

3.1 Electronics

The principal upgrade regards the substitution of the PMTs with Silicon PhotoMultipliers (SiPMs), one for each fiber. The ASD RGB-SiPMs provided by AdvanSiD[3] have a 1×1 mm^2 active area completely covered with a transparent epoxy layer and they are located inside a SMD plastic package of 2.03 x 2.48 x 1.30 mm^3. The active area is composed by 625 micro-cells of 40 μm each one. The SiPMs are based on the "N-on-P" silicon technology for detection of Red, Green, and Blue light. They have a peak efficiency at 550 nm, with a detection spectrum extending from 350 nm to 900 nm. The use of the SiPMs guarantees a better coupling with the fibers and allows to have a more compact system. The SiPMs can operate in high magnetic field, they do not need high voltage and moreover the chosen model front-end readout scheme and radiation tolerance have already been tested by some members of the collaboration [19]. A better efficiency uniformity is then expected. SiPMs are mounted on new electronic boards that are then connected with the front-end boards (FEB) equipped with a MAROC3 ASIC, already in use. The advantage of this system is that these new boards can be located inside the frame and each SiPM connected with a single fiber.

3.2 Mechanical support

In the past configuration, bars were surrounded by an aluminium frame and located between two black plastic panels that hooked with the frame to create a unique and closed system. In this way, bars and fibers were not fixed in a tight way inside the frame. We improved the mechanical support, making the bars self-supporting. On the two extremities of the bars, where the fibers exit from the plastic scintillator, a pair of L-shaped aluminium extrusions were positioned one on the top and one on the bottom, tied together and hinged to the external aluminium frame. This avoids movements in the 3 directions and makes the system of the bars plus the external frame independent from the 2 black covers. A picture of the mechanical supports is shown in figure 3.

3.3 Fibers coupling

As previously described each fiber must be connected with a single SiPM that is located on the new electronic board. An accurate coupling between the 1x1 mm^2 active square area of the SiPM and the circular 1 mm diameter transmitting area of the fiber is necessary to avoid any loss of signal. Since the position of the SiPM on the board is fixed and the nominal mechanical tolerance of the black package is ± 0.15 mm (both on length and width), we studied the position of the active area with respect to the housing.

[3]https://advansid.com/home

https://doi.org/10.1051/epjconf/202226201013

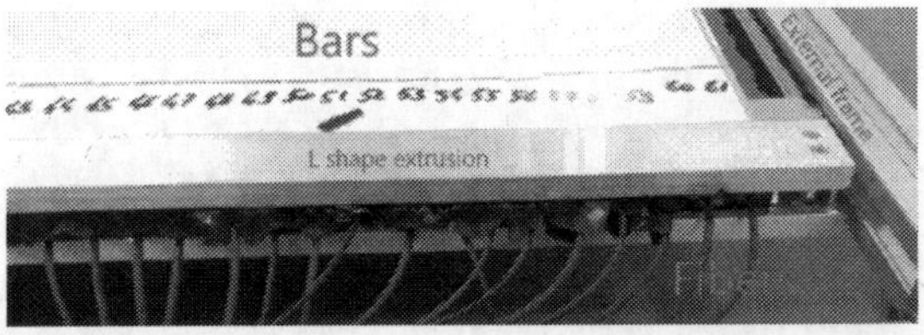

Figure 3: New mechanical support. The scintillating bars are held by 2 L-shaped aluminium extrusions tied together and hinged to the external aluminium frame, to obtain a self-supporting system.

Figure 4: Picture of one SiPM. The orthogonal Cartesian system has the origin in the bottom-left vertex of the active area and the four vertices are labeled with points A,B,C,D [20].

	$\bar{x}$ [mm]	Std. Dev x [mm]
A	−0.50	0.04
B	1.98	0.04
C	1.97	0.03
D	−0.51	0.03

	$\bar{y}$ [mm]	Std. Dev y [mm]
A	−0.52	0.02
B	−0.51	0.02
C	1.58	0.03
D	1.56	0.02

Table 1: Average values of the x and y coordinates with relative errors together with the standard deviation of the measurements for the 4 vertices of the external package for the 98 SiPMs.

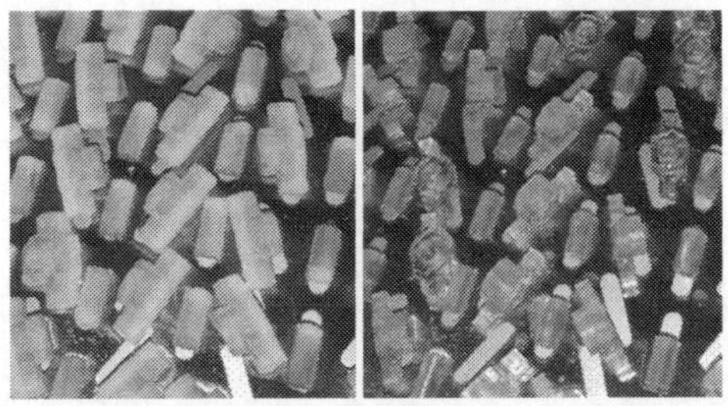

Figure 5: Pictures of the two parts of the couplers, produced with the Project 2500 Pro machine that works with material jetting technology. HT250, a photocurable resin, was selected as part material, wax for support material. After the production, the parts underwent a post-process (water bath, immersion in mineral oil and ultrasonic cleaning) to eliminate the support material. The couplers before (left side) and after (right side) the water bath are shown.

A photo camera Canon with MACRO lens was used to take high-resolution photos of the upper surface of each SiPM. The relative position between the plastic support and the active area was obtained by analyzing the images with an online tool [20] that allows to measure distances and get points coordinates precisely once a reference system is defined. Due to the difficulty to arrange each SiPM always in the same position with respect to the camera, we identified an orthogonal Cartesian system with the origin in one of the vertices of the active area and the axes along two sides of the active square as reported in figure 4. In this way, even if the coordinate system is different for each image, our process does not depend on the possible mismatch position of the SiPMs. After the reference frame is defined, we identified the coordinates of the four vertices of the external package with respect to the axes origin as displayed in figure 4. We repeated this process for 98 SiPMs, randomly chosen, and we calculated the average values of the coordinates and their standard deviations. The results are listed in table 1. We can observe that the standard deviation is of the order of few 10^{-2} mm.

Then we focused on the planning of the fiber-SiPM coupling. We used a 3D printer to mould a coupler in resin composed by two parts: the first one that is going to be fixed on the board and the second that has to be glued to the fiber. At the end they will be connected to finalize the coupling. They were produced with the Project 2500 Pro machine (3D System, South Caroline, US) that works with material jetting technology; this technology is an inkjet printing process that uses piezo printhead technology to deposit photocurable plastic resin droplets layer by layer. HT250, a photocurable resin provided by 3D System, was selected as the part material, while wax for the support material. After the production, the parts underwent a post process (water bath, immersion in mineral oil and ultrasonic cleaning) to eliminate the support material. In figure 5 the couplers before (left side) and after (right side) the water bath are shown.

The component secured on the board is created with a rectangular hole where the SiPM can be housed. In order to know the correct dimension of the rectangle, we used the data obtained from the position of the vertex to determine the lengths of the 4 sides. The result of the analysis on 98 SiPMs, shows that the average values are always compatible with the nominal ones within the mechanical tolerance, even if the average values for the two short sides (BC and DA) are 3% higher than the nominal lengths. This analysis has allowed us to produce the couplers with the correct dimension for housing the SiPMs.

The other component, the one that is fixed to the fiber, consists in a cylinder where the fiber is stringed. The coupling of the two components with the board is illustrated in figure 6.

Figure 6: 3D scheme of the coupling between the two resin couplers (the ocher and the red parts) and the electronic board (green part) on which the SiPMs are fixed.

All the fibers were cut, inserted in the cylinders, glued with epoxy and then polished. In figure 7 the couplers with the housing for the SiPMs, fixed to the board and the ones coupled with the fibers are shown. At this point, the cylinders are inserted in the board component and the fibers and the SiPMs are coupled.

Figure 7: Pictures of the couplers. The first component with the housing for the SiPMs is fixed to the board and the second one is glued with fibers. The cylinders must be slotted in the board component and the fibers and the SiPMs can be coupled.

4 Preliminary test

Preliminary measurements are now in progress with cosmic-rays. As shown in the upper part of figure 8, two plastic scintillators and two silicon microstrip XY detectors are located under one of the plane of the scintillating bars. The coincidence between the two scintillators provides the trigger for the data acquisition system and the microstrip detectors are used to track the cosmic rays. Extrapolating the hit position of each track on the bars plane and recording their signals it is possible to characterize each channel in terms of detection efficiency and uniformity of the light yield. Many weeks of data taking are needed to collect enough statistic and the analysis is still in progress. Nevertheless, the already acquired data (e.g. the hit profile of one plane is shown in the lower part of figure 8) give a feedback on the uniformity of the optical fibers-SiPM coupling and small interventions to correct defects are in progress.

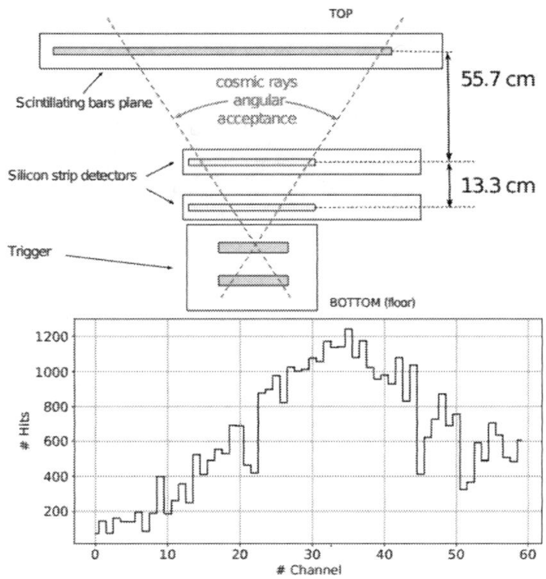

Figure 8: Upper part: the experimental setup for cosmic data taking campaign where a scintillating bar plane is put under test thanks to two plastic scintillators and two silicon microstrip detectors as a reference. Lower part: a hit profile of the cosmic rays on one plane. The plot shape is determined by the angular distribution of the cosmic rays and the geometrical acceptance of the trigger detectors, whereas local non uniformities are presumably due to bad fiber-SiPM couplings or a mechanical pull of the electronic board hosting the SiPMs. In both cases, they could be corrected by simple mechanical adjustments.

5 Conclusion

The upgrade of the scintillating bars detectors for the ASACUSA experiment has been presented. It mainly consists in the replacement of PhotoMultiplier tubes with Silicon PhotoMultipliers, that allow a better coupling with fibers and better performances in high magnetic field. The mechanical support has been modified to make the bars self-supported and resin couplers have been designed and produced for connecting a fiber with a SiPM. Performance and efficiency tests are now in progress.

References

[1] H. Aghai-Khozani, A. Bianconi, M. Corradini, R. Hayano, M. Hori, M. Leali, E. Lodi Rizzini, V. Mascagna, Y. Murakami, M. Prest et al., Nuclear Physics A **970**, 366 (2018)

[2] K. Todoroki, D. Barna, R. Hayano, H. Aghai-Khozani, A. Sótér, M. Corradini, M. Leali, E. Lodi-Rizzini, V. Mascagna, L. Venturelli et al., Nuclear Instruments and Methods in Physics Research Section A: Accelerators, Spectrometers, Detectors and Associated Equipment **835**, 110 (2016)

[3] H. Aghai-Khozani, D. Barna, M. Corradini, R. Hayano, M. Hori, T. Kobayashi, M. Leali, E. Lodi-Rizzini, V. Mascagna, M. Prest et al., The European Physical Journal Plus **127**, 125 (2012)

[4] H. Aghai-Khozani, D. Barna, M. Corradini, D. De Salvador, R. Hayano, M. Hori, T. Kobayashi, M. Leali, E. Lodi-Rizzini, V. Mascagna et al., Hyperfine Interactions **234**, 85 (2015)

[5] H. Aghai-Khozani, D. Barna, M. Corradini, D. De Salvador, R. Hayano, M. Hori, M. Leali, E. Lodi-Rizzini, V. Mascagna, M. Prest et al., Nuclear Physics A **1009**, 122170 (2021)

[6] Y. Enomoto, N. Kuroda, K. Michishio, C.H. Kim, H. Higaki, Y. Nagata, Y. Kanai, H.A. Torii, M. Corradini, M. Leali et al., Phys. Rev. Lett. **105**, 243401 (2010)

[7] C. Malbrunot, C. Amsler, S. Arguedas Cuendis, H. Breuker, P. Dupre, M. Fleck, H. Higaki, Y. Kanai, B. Kolbinger, N. Kuroda et al., Philosophical Transactions of the Royal Society A: Mathematical, Physical and Engineering Sciences **376**, 20170273 (2018)

[8] V. Chohan, C. Alanzeau, M.E. Angoletta, J. Baillie, D. Barna, W. Bartmann, P. Belochitskii, J. Borburgh, H. Breuker, F. Butin et al., *Extra Low ENergy Antiproton (ELENA) ring and its Transfer Lines: Design Report*, CERN Yellow Reports: Monographs (CERN, Geneva, 2014), `https://cds.cern.ch/record/1694484`

[9] Y. Nagata, N. Kuroda, P. Dupre, B. Radics, M. Tajima, A.A. Capon, M. Diermaier, C. Kaga, B. Kolbinger, M. Leali et al., Proceedings of the 12th International Conference on Low Energy Antiproton Physics (LEAP2016) (2017)

[10] A. Mohri, H. Higaki, H. Tanaka, Y. Yamazawa, M. Aoyagi, T. Yuyama, T. Michishita, Japanese Journal of Applied Physics **37**, 664 (1998)

[11] C. Sauerzopf, A.A. Capon, M. Diermaier, P. Dupré, Y. Higashi, C. Kaga, B. Kolbinger, M. Leali, S. Lehner, E.L. Rizzini et al., Hyperfine Interactions **237**, 103 (2016)

[12] N. Kuroda, S. Ulmer, D.J. Murtagh, S.V. Gorp, Y. Nagata, M. Diermaier, S. Federmann, C.M. M. Leali, V.Mascagna, O. Massiczek et al., Nature Communications **5**, 3089 (2014)

[13] C. Malbrunot, M. Diermaier, M. Simon, C. Amsler, S. Arguedas Cuendis, H. Breuker, C. Evans, M. Fleck, B. Kolbinger, A. Lanz et al., Nuclear Instruments and Methods in Physics Research Section A: Accelerators, Spectrometers, Detectors and Associated Equipment **935**, 110 (2019)

[14] E. Widmann, C. Amsler, S. Arguedas Cuendis, H. Breuker, M. Diermaier, P. Dupré, C. Evans, M. Fleck, A. Gligorova, H. Higaki et al., Hyperfine Interactions **240**, 5 (2018)

[15] Y. Nagata, N. Kuroda, M. Ohtsuka, M. Leali, E. Lodi-Rizzini, V. Mascagna, M. Tajima, H. Torii, N. Zurlo, Y. Matsuda et al., Nuclear Instruments and Methods in Physics Research Section A: Accelerators, Spectrometers, Detectors and Associated Equipment **840**, 153 (2016)

[16] B. Kolbinger, C. Amsler, H. Breuker, M. Diermaier, P. Dupré, M. Fleck, A. Gligorova, H. Higaki, Y. Kanai, T. Kobayashi et al., EPJ Web of Conferences **181** (2018)

[17] A. Pla-Dalmau, A.D. Bross, K.L. Mellott, Nuclear Instruments and Methods in Physics Research Section A: Accelerators, Spectrometers, Detectors and Associated Equipment **466**, 482 (2001)

[18] M. Corradini, M. Leali, E. Lodi-Rizzini, V. Mascagna, V. Prest, L. Vallazza, L. Venturelli, Hyperfine Interactions **233**, 53 (2015)

[19] F. Acerbi, G. Ballerini, A. Berra, C. Brizzolari, G. Brunetti, M. Catanesi, S. Cecchini, F. Cindolo, A. Coffani, G. Collazuol et al., Journal of Instrumentation **14**, P02029 (2019)

[20] A. Rohatgi, *Webplotdigitizer: Version 4.5* (2021), `https://automeris.io/WebPlotDigitizer`

BRAND – exploring transverse polarization of electrons emitted in neutron decay

Feasibility demonstration experiment

K. Bodek[1,*], *J.* Choi[5], *L.* De Keukeleere[3], *K.* Dhanmeher[2], *M.* Engler[6], *G.* Gupta[1], *A.* Kozela[2], *K.* Lojek[1], *K.* Pysz[2], *D.* Ries[6], *D.* Rozpedzik[1], *N.* Severijns[3], *T.* Soldner[4], *N.* Yazdandoost[6], *A.R.* Young[5], and *J.* Zejma[1]

[1]M. Smoluchowski Institute of Physics, Jagiellonian University, Krakow, Poland
[2]H. Niewodniczanski Institute of Nuclear Physics, Polish Academy of Sciences, Krakow, Poland
[3]Institute of Nuclear and Radiation Physics, KU Leuven, Belgium
[4]Institut Laue-Langevin, Grenoble, France
[5]Dep. of Physics and Astronomy, North Carolina State University, Raleigh, USA
[6]Dep. of Chemistry, TRIGA site, J. Gutenberg University, Mainz, Germany

Abstract. Neutron and nuclear beta decay correlation coefficients are sensitive to the exotic scalar and tensor interactions that are not included in the Standard Model (SM). The proposed experiment BRAND will measure simultaneously seven neutron correlation coefficients: H, L, N, R, S, U and V that depend on the transverse electron polarization – a quantity which vanishes in the SM. Five of these correlations: H, L, S, U and V were never attempted experimentally before. The expected impact of the proposed experiment is comparable to that of frequently measured "traditional" correlation coefficients (a, b, A, B, D) but offers completely different systematics and additional sensitivity to imaginary parts of the scalar and tensor couplings. In order to demonstrate the feasibility of the challenging techniques such as the event-by-event decay kinematics reconstruction together with the electron polarimetry a test setup was installed at the cold neutron beam line PF1B at the Laue-Langevin Institute, Grenoble, France. In this contribution, the results of the first run as well as plans for the run in Autumn 2021 will be discussed.

1 Introduction

Among the empirical foundations of the electroweak standard model (SM) the assumptions of maximal parity violation, vector and axial-vector character and massless neutrinos are directly linked to the experiments performed with the nuclear and neutron beta decay. The beta decay theory has been firmly established about six decades ago and became a part of the SM. Nevertheless, despite the great success of the SM, many open questions remain such as the origin of parity violation, the hierarchy of fermion masses, the number of particle generations, the mechanism of CP violation, the worrying large number of parameters of the theory, etc. Nuclear beta decay experiments test weakly interacting systems and free neutron

*e-mail: kazimierz.bodek@uj.edu.pl

decay plays a particular role: due to its simplicity it is free of complications connected with the nuclear and atomic structure.

The neutron beta decay is an allowed and mixed transition. Contrary to the nuclear mixed beta transitions, the Fermi and Gamow-Teller contributions to the neutron decay rate and energy-angle distribution are known accurately since the corresponding matrix elements are not affected by nuclear structure uncertainties. In the SM, the neutron beta decay is described by only two parameters: the V_{ud} element of the Cabibo-Kobayashi-Maskawa mixing matrix and the ratio of the axial vector to vector charges, $\lambda = g_A/g_V$. The large number of the experimental observables depending on the momenta and spins of the involved particles makes the situation highly redundant and attractive for precision experiments aiming at the improvement of the SM and searching for effects which cannot be explained by the SM. For instance, the SM predictions of time-reversal violation for weak decays of systems built up of u and d quarks are by 7 to 10 orders of magnitude lower than the experimental accuracies attainable at present. The exploration of this intriguing window is mitigated by the accuracy of theoretically calculated electromagnetic (EM) corrections which presently achieve the level slightly better than 10^{-4} [1, 2].

2 Transverse electron polarization in neutron decay

Generally, in the Standard Model, the electrons emitted in beta decay are longitudinally polarized which reflects the parity violating V-A structure of the weak interaction. The departure of the polarization vector from a strict collinearity with electron momentum can be caused by electromagnetic effects, recoil order corrections (induced couplings) or exotic scalar and tensor interactions. Thus, provided that the electromagnetic and recoil order corrections are under control, the transverse electron polarization is an ideal observable for searches of physics beyond the Standard Model.

The transverse electron polarization is reflected in the distribution of the decay products via a number of the correlation coefficients relating it to other vectors characterizing the system, the most important being the electron and antineutrino momenta $\mathbf{p}_e$, $\mathbf{p}_{\bar{\nu}}$ and the neutron spin $\mathbf{J}$. The corresponding formula limited to the lowest order terms can be found in the classical papers [3–5]. Dropping out all the terms not depending explicitly on the transverse components of the electron polarization and retaining five exceptions: a, b, A, B and D ("classical" correlations) one arrives at:

$$\begin{aligned}\omega(E_e,\Omega_e,\Omega_{\bar{\nu}}) \;\propto\; 1 \;+ \\ a\,\frac{\mathbf{p}_e\cdot\mathbf{p}_{\bar{\nu}}}{E_eE_{\bar{\nu}}} + b\,\frac{m_e}{E_e} + \frac{\langle\mathbf{J}\rangle}{J}\cdot\left[A\,\frac{\mathbf{p}_e}{E_e} + B\,\frac{\mathbf{p}_{\bar{\nu}}}{E_{\bar{\nu}}} + D\,\frac{\mathbf{p}_e\times\mathbf{p}_{\bar{\nu}}}{E_eE_{\bar{\nu}}}\right] + \\ \sigma_\perp\cdot\left[H\,\frac{\mathbf{p}_{\bar{\nu}}}{E_{\bar{\nu}}} + L\,\frac{\mathbf{p}_e\times\mathbf{p}_{\bar{\nu}}}{E_eE_{\bar{\nu}}} + N\,\frac{\langle\mathbf{J}\rangle}{J} + R\,\frac{\langle\mathbf{J}\rangle\times\mathbf{p}_e}{J\,E_e} + S\,\frac{\langle\mathbf{J}\rangle}{J}\,\frac{\mathbf{p}_e\cdot\mathbf{p}_{\bar{\nu}}}{E_eE_{\bar{\nu}}} + U\,\mathbf{p}_{\bar{\nu}}\,\frac{\langle\mathbf{J}\rangle\cdot\mathbf{p}_e}{J\,E_eE_{\bar{\nu}}} + V\,\frac{\mathbf{p}_{\bar{\nu}}\times\langle\mathbf{J}\rangle}{J\,E_{\bar{\nu}}}\right],\end{aligned} \tag{1}$$

where $\boldsymbol{\sigma}_\perp$ represents a unit vector perpendicular to the electron momentum $\mathbf{p}_e$ and $J = |\mathbf{J}|$. In the infinite neutron mass approximation (no recoil), making usual assumptions for the Standard Model: $C_V = C'_V = 1$ and $\lambda \equiv C_A = C'_A = -1.272$ [6], and neglecting the contributions quadratic (and higher order) in C_S, C'_S, C_T, C'_T one can express all the correlation coefficients from Eq. (1) (called here X) as combinations of the real and imaginary parts of the scalar, $\mathfrak{S}$, and tensor, $\mathfrak{T}$, normalized exotic couplings:

$$X = X_{\rm SM} + X_{\rm EM} + c^X_{{\rm Re}S}\,{\rm Re}(\mathfrak{S}) + c^X_{{\rm Re}T}\,{\rm Re}(\mathfrak{T}) + c^X_{{\rm Im}S}\,{\rm Im}(\mathfrak{S}) + c^X_{{\rm Im}T}\,{\rm Im}(\mathfrak{T}) \tag{2}$$

with

$$\mathfrak{S} \equiv \frac{C_S + C'_S}{C_V}, \qquad \mathfrak{T} \equiv \frac{C_T + C'_T}{C_A}. \tag{3}$$

The coefficients c in this expression are functions of λ and kinematical quantities. Table 1 in Ref. [7] summarizes their values calculated with the kinematical factors averaged over the electron spectrum in the kinetic energy range 200–783 keV.

The coefficients relating the transverse electron polarization to $\mathbf{p}_e$, $\mathbf{p}_{\bar{\nu}}$ and $\mathbf{J}$ have several interesting features. They vanish for the SM weak interaction, reveal variable size of the EM contributions and are measurable using known techniques. And, last but not least, the dependence on real and imaginary parts of the scalar and tensor couplings alternates exclusively from one correlation coefficient to another with varying sensitivity coefficients.

3 BRAND project

The experiment reported in Refs. [9–12] has clearly proven that the transverse electron polarization of electrons emitted in the decay of cold neutrons can be accurately measured with Mott scattering. The applied technique of electron tracking and reconstruction of the scattering vertices is a powerful tool for reduction of background typical in experiments with intense neutron beams. The rich experience and quantitative information gained in that pioneering experiment allow for a realistic planning of the next steps.

The proposed experiment will measure the transverse electron polarization expressed in terms of 7 correlation coefficients (H, L, N, R, S, U, V) as compared to two (N, R) accessible in the former setup [9–12]. The simultaneous measurements of the coefficients a, A, B and D will help to keep systematics under control. It will be intriguing to compare their values with those from the specialized experiments [13–20] since our experimental technique is completely different.

The measurement of momenta of electrons and protons in coincidence will allow for reconstruction of the antineutrino momentum and will provide access to the terms dependent on this quantity. The strategy of the proposed experiment is to reconstruct event-by-event the decay kinematics and identify the decay origin. Such a condition will reduce the effects due to large size of the decay source, as well as scattering effects and will significantly suppress background. In particular, confining the three-body decay kinematics by the measured electron energy and relative $e-p$ angle one realizes that the proton energy and thus the proton time-of-flight must choose between two discrete (and known) values.

The key features of the proposed setup are: (i) Efficient cylindrical detector geometry. (ii) Electron tracking in a multi-wire drift chamber (MWDC) with hexagonal geometry of cells with signal readout at both wire ends operated in He/isobutane gas mixture. (iii) Detection of both direct and Mott-scattered electrons in plastic scintillator hodoscopes. (iv) Conversion of protons (accelerated to 20–30 keV) into bunches of secondary electrons ejected from a thin LiF layer [21]. Acceleration and subsequent detection of ejected electrons in a thin (~25 μm) plastic scintillator with position sensitivity. (v) Reduction of geometry misalignment related systematic effects by the use of advanced polarimetry techniques such as periodic neutron spin flip during data taking and evaluation of asymmetry and super-ratio in data analysis.

Background contributions will be subtracted using data collected in several regimes (beam-on, beam-off, Mott target-on, Mott target-off). In order to deduce necessary corrections and asses associated systematic errors, mapping of the spin guiding field and detector efficiency will be carried out in a series of calibration experiments.

A cross section of the proposed cylindrical Mott polarimeter for decay electrons and accompanying detectors for recoil protons is sketched in Figure 1. Decay electrons reach the Mott target made a of 1–2 μm thick layer of high Z element (lead or depleted uranium) deposited on a thin Mylar substrate. The decay electrons can be either backscattered or pass the foil and be detected in the outer plastic scintillator hodoscope located close to it. The Mott-scattered electrons are registered in either of the inner hodoscopes as shown in Figure 1. The electron hodoscopes will be made of plastic scintillator bars readout by photomultiplier tubes attached to both bar ends. They will operate at ambient pressure outside of the vacuum chamber.

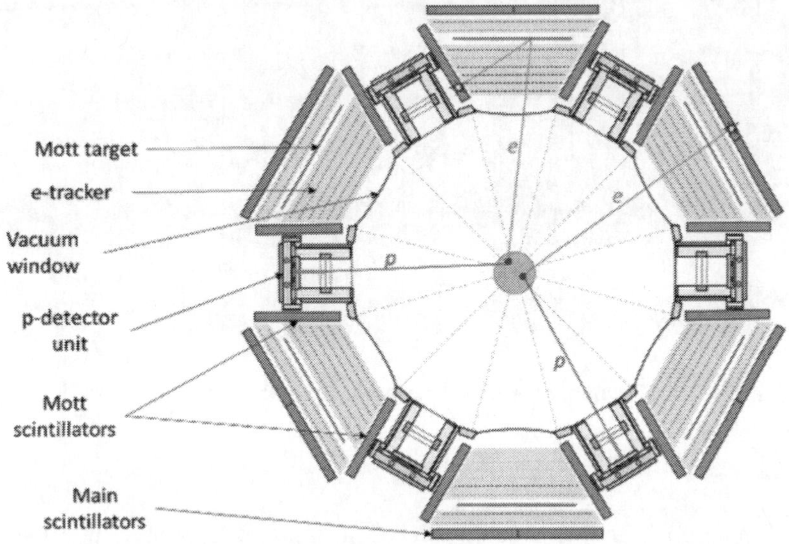

Figure 1. (Color on-line) Sketch of the proposed experimental setup in the cylindrical geometry.

After leaving the free drift zone, protons are accelerated by electric field towards a $p - e$ conversion foil consisting of a 20 nm thick LiF layer deposited onto a thin (100 nm) fluorinated-polyamide film [21]. The located in vacuum "proton" detectors will register bunches of up to 20 secondary electrons ejected from that foil. Acceleration and subsequent detection of ejected electrons is implemented using a thin ($\approx$25 μm) plastic scintillator film deposited onto light guides. The scintillation light will be registered by SiPM sensors attached along the light guide. The hit position information will be deduced from the distribution of the light signal registered by SiPM sensors close to the light flash source. The "proton" detector will be able to discriminate electrons from beta decay with energy higher than 30 keV since they deposit in it less than 30 keV while the accumulated signal from the electron bunch following the $p - e$ conversion corresponds to 100–500 keV.

3.1 Demonstration apparatus

The proposed BRAND experiment assumes several difficult techniques to be applied simultaneously. Among them are large area thin vacuum windows, precision low-mass, low-Z electron tracker and large acceptance recoil proton detector made of low Z materials (required to minimize backscattering of charged particles). In order to check the proposed solutions, in September 2020, a short test measurement (5 days) with the use of prototype BRAND apparatus (called BRAND-0 setup) was performed at the neutron facility PF1B of the Laue-Langevin Institute (ILL), Grenoble. Polarized cold neutron beam with a flux of 6.3×10^8

n/cm^2/s and cross-section of 6×6 cm^2 was used for this purpose. Fig. 2 shows a sketch of the BRAND-0 setup.

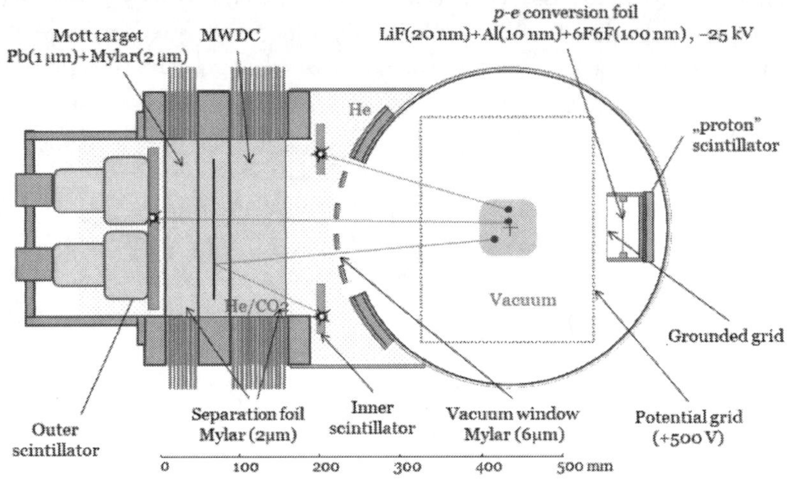

Figure 2. (Color on-line) Layout of the test experiment BRAND-0: Cross section in the plane perpendicular to the beam axis.

The cold neutron beam was transported along axis of a cylindrical vacuum chamber. A fraction of electrons from the neutron decay could leave the decay chamber through the thin side windows ($\approx$10 μm thick aluminized Mylar) and were registered in the electron detector consisting of tracking and plastic scintillator parts. Electron tracks were measured in e low density helium based (He:CO_2/85:15 + trace of ethanol vapor)[1] gaseous detector operated at ambient pressure. It was built as an 8 layer drift chamber of hexagonal cells. The track position in XY-plane was extracted from the measured drift time of ionization electrons whereas the position along the wire length (YZ-plane) was obtained by means of the charge division technique. The prototype electron tracker is a copy of the miniBETA spectrometer [22, 23]. Its properties are optimized for the energy range of electrons emitted in β-decays.

Electron tracking capability, a thin lead scatterer and additional scintillators at backward angles made this system an electron spin analyzer utilizing the Mott scattering. The Mott target (2 μm of Pb evaporated on 4 μm thick Mylar foil) was installed inside the tracker. Electrons passing the tracker and back-scattered by means of Mott process could be identified by characteristic track topology (called V-track). The energy of scattered electrons was measured in two, rectangular 10 mm thick plastic scintillators symmetrically positioned near to the tracker entrance window. The electrons transmitted through the Mott foil without significant scattering were registered in a circular 10 mm thick plastic scintillator installed behind the tracker. All scintillator detectors triggered DAQ, measured energy of registered electrons and provided the start signal for drift time measurement.

In the BRAND-0 setup, the proton detector prototype together with preamplifier boards was installed inside the decay chamber. Heat removal was accomplished with an ordinary cooling circuit. Stable temperature of about 12 ^{o}C was provided during measurements. Laboratory tests delivered the hit position resolution of such a system better than 3 mm [24].

[1]Less efficient but non-flammable gas mixture was used in the test due to safety regulations at ILL.

3.2 Analysis of test data

Accurate identification and precise tracking of electrons is crucial in the BRAND experiment since the transverse electron polarization is deduced from angular distribution in the Mott scattering. In the preliminary analysis of the collected data the average spacial resolution of MWDC in XY-projection deduced from drift time is about 0.4 mm (Fig. 3a) while the charge division technique provides 5 mm resolution in Z direction (Fig. 3b). An example V-track event topology is presented in Fig. 3c while the distribution of the reconstructed V-track vertex position is shown in Fig. 3d. A pronounced peak seen in this distribution at $X = 372$ mm points to the position of the Mott target. The scattering vertices in the Mott target are well separated from background ones attributed mainly to scattering on wires.

The obtained vertex position resolution of about 24 mm agrees with expectation. It should be pointed out that the prototype electron tracker fulfils the requirements posed by the BRAND project. The single cell position resolution deduced from drift time translates to the uncertainty of reconstructed angles in XY-projection. These, in turn, translate to the Mott scattering vertex position resolution critical for selection of events that belong to the Mott target. It is expected that the single cell position resolution should not exceed 0.7 mm. The vertex position resolution achieved in the test is roughly a factor of two worse than the required one. The difference with the expectation can be explained by the range of scattering angles accessible in the test that preferred acute V-track angles (20° – 40°). In the ultimate detector geometry, the V-track angles will range between 50° – 80°.

The required Z-position resolution in the ultimate setup is estimated to about 25 mm. Taking into account wire length (reduced by a factor of five as compared to the ultimate solution), the resolution of 5 mm achieved in the test fulfills exactly the requirement.

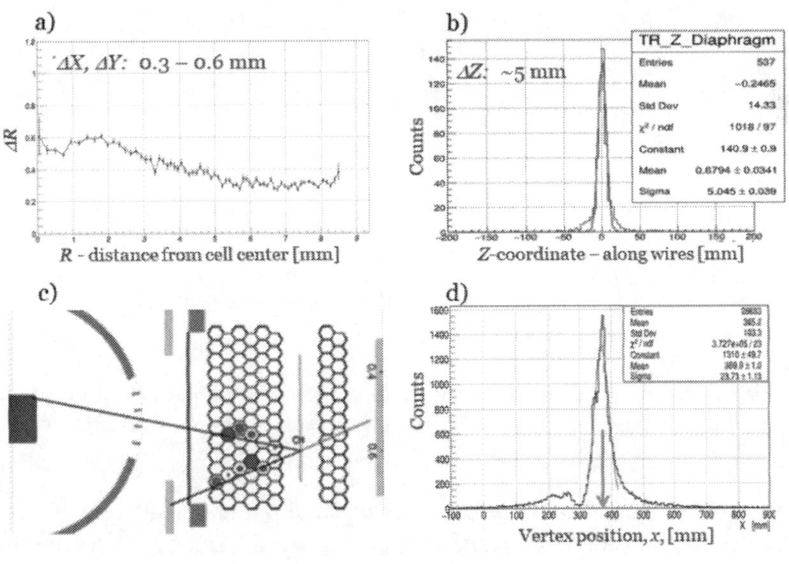

Figure 3. (Color on-line) Preliminary results achieved with the test setup BRAND-0. a) Position resolution extracted from drift time. b) Position resolution extracted from charge division technique. c) V-track event topology. d) Distribution of the reconstructed V-track vertex position (see text for more details).

The very short beam time available to the test and incomplete functionality of the trigger board and SiPM control system allowed only for initial performance test of the first prototype proton detector. Equipped with a single 2 inch diameter circular converter foil the detector was tested in the self-triggering mode. The electric field configuration was optimized using COMSOL simulations. The goal was to collect as many as possible recoil protons emitted in the beam in order to increase the detection rate in a small detector (Fig. 4, left panel). In addition, the completed ad hoc high voltage system could sustain only about 16 kV so that the detector transmission was far from optimum. Nevertheless, we observed sensible behavior of rates and pulse height distribution. The plotted pulse height spectra reveal growing excess of counts and larger pulse height with increasing acceleration potential as shown in Fig. 4, right panel. Therein the line corresponding to −6 kV can be interpreted as the noise contribution. At this voltage the multiplicity of secondary electrons is significantly reduced [21] and the signal is obviously below the discrimination threshold. The increasing cage potential leads to increase of the proton acceleration potential and, consequently, to higher multiplicity of secondary electrons and larger signals as seen from Fig. 4, lower-right panel.

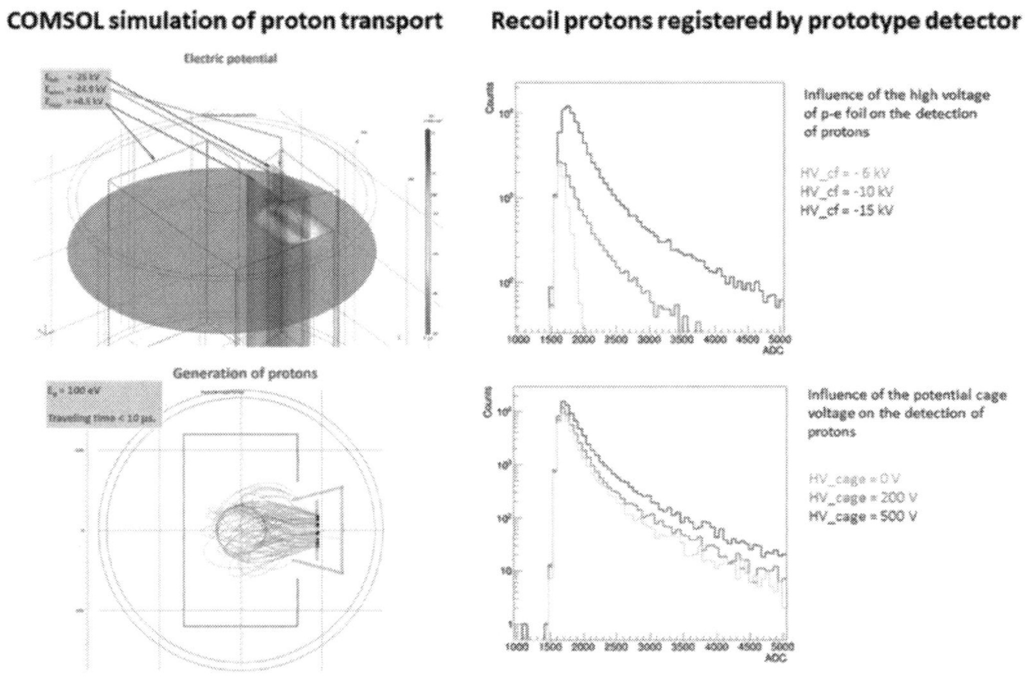

Figure 4. (Color on-line) Proton detector prototype in the self-triggering mode. Left panel: COMSOL simulation of proton transport. Right panel: Growing excess of counts with increasing acceleration potential.

Contrary to positive test results for the electron tracking and Mott polarimetry, it is clear that the proton detector prototype must be further improved. Among others the SiPM control and the gain calibration system (using alpha source) must be implemented before the hit position reconstruction will be possible.

4 BRAND-1 setup

Encouraging results from the first on-beam test of the detector prototypes led to the second experiment with improved components and the setup geometry closer to that of the ultimate experiment. Therein the Mott detectors are embedded in the MWDC. One of the important goals for measurements with the BRAND-1 setup is to produce data with the complete chain of detectors, front end electronics and DAQ. Moreover, the neutron spin guiding field and periodic spin flip procedure is also implemented such that a full analysis of data and extraction of the correlation coefficients will be possible. Fig. 5 shows a sketch of the BRAND-1 setup. The apparatus was installed on the PF1B beam line at ILL, Grenoble and collected data within a four week long run starting in September 2021. Presently, the collected data are subject of intensive analysis.

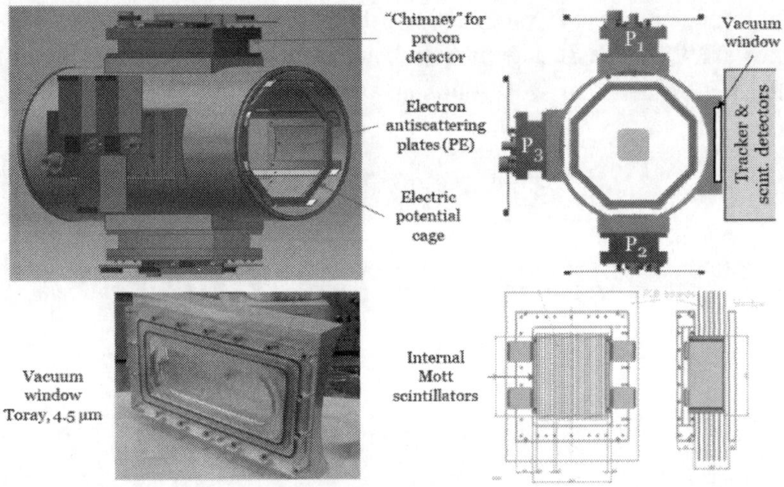

Figure 5. (Color on-line) BRAND-1 setup. Upper panel: Decay chamber with compartments for three proton detectors. Lower-left panel: Kevlar thread reinforced vacuum window prototype. Lower-right panel: Sketch of Mott detectors embedded in the electron tracker.

Acknowledgments: This work has been supported by the National Science Center, Poland, under the grant No. UMO-2018/29/B/ST2/02505. NCSU co-authors acknowledge the supporting grants from NSF: PHY-1914133 and DOE: DE-FG02-ER41042.

References

[1] A.N. Ivanov et al., Phys. Rev. C 95, 055502 (2017) and references therein.
[2] A.N. Ivanov et al., Phys. Rev. C 98, 035503 (2018) and references therein.
[3] J. Jackson et al., Phys. Rev. 106, 517 (1957).
[4] J. Jackson et al., Nucl. Phys. 4, 206 (1957).
[5] M.E. Ebel et al., Nucl. Phys. 4, 213, (1957).
[6] P.A. Zyla et al. (Particle Data Group), Prog. Theor. Exp. Phys. 2020, 083C01 (2020) and 2021 update.
[7] K. Bodek et al., EPJ Web of Conf. 219, 04001 (2019).
[8] V. Khachatryan, et al. CERN Rep. CMS-PAS-EXO-12-060, (2013).
[9] A. Kozela et al., Phys. Rev. Lett. 102, 172301 (2009).
[10] G. Ban et al., Nucl. Instrum. Methods Phys. Res., Sect. A 565, 711 (2006).
[11] K. Bodek et al., Physics Procedia 17, 30 (2011).
[12] A. Kozela et al., Phys. Rev. C 85, 045501 (2012).
[13] W. S. Wilburn et al., J. Res. NIST 110, 389 (2005).

[14] B. Maerkisch et al., Nucl. Instr. Meth. Phys. Res. A 611, 216 (2009).
[15] D. Dubbers et al., Nucl. Instr. Meth. Phys. Res. A 596, 238 (2008).
[16] O. Zimmer et al., Nucl. Instr. Meth. Phys. Res. A 440, 548 (2000).
[17] F.E. Wietfeldt et al., Nucl. Instr. Meth. Phys. Res. A 611, 207 (2009).
[18] B. Tipton et al., AIP Conf. Proc. 539, 286 (2000).
[19] R. Alarcon et al., URL http://nab.phys.virginia.edu/ nab_loi.pdf, Letter of intent (2005)
[20] H.P. Mumm et al., Rev. Sci. Instr. 75, 5343 (2004).
[21] S. Hoedl et al., J. Appl. Phys. 99, 084904 (2006).
[22] M. Perkowski et al., Acta Phys. Pol. 49 (2018), p. 261.
[23] M. Perkowski. PhD Thesis, Jagiellonian University and KUL, Krakow/Leuven, 2020.
[24] M. Kolodziej, Master Thesis, Jagiellonian University, Krakow, 2020.

https://doi.org/10.1051/epjconf/202226201015

Prospects for a neutron EDM measurement with an advanced ultracold neutron source at TRIUMF

Takashi Higuchi[1,*] on behalf of the TUCAN collaboration

[1]Research Center for Nuclear Physics, Osaka University,
10-1 Mihogaoka, Ibaraki, Osaka 567-0047, Japan

Abstract. The TUCAN (TRIUMF Ultra-Cold Advanced Neutron) collaboration is building a new high-intensity ultracold neutron (UCN) source at TRIUMF with the aim of measuring the neutron electric dipole moment (nEDM) with unprecedented precision. TUCAN employs a spallation-driven superthermal UCN production scheme which has been demonstrated with a prototype UCN source. In this article, recent progress on the major components of a new upgraded UCN source and developments in thee context of installation of the nEDM spectrometer are reported.

1 Introduction

1.1 Background

Tests of discrete symmetries have occupied an important place in modern particle physics since parity (P) violation in the weak interaction was first discovered [1, 2]. In 1967, Sakharov proposed a scenario for explaining the baryon-antibaryon asymmetry of the universe based on a few fundamental conditions, including Charge-Parity (CP) violation [3]. The baryon asymmetry η of the current universe is defined as

$$\eta = \frac{n_B - n_{\bar{B}}}{n_\gamma}, \tag{1}$$

with $n_B, n_{\bar{B}}$ and n_γ being the number densities of baryonic matter, antibaryonic matter and photons, respectively. Based on CP violation of the Standard Model (SM), η is predicted to be $\eta_{\rm SM}\sim 10^{-20}$ [4]. However, $\eta_{\rm obs}\sim 10^{-10}$ is found from observation of the cosmic microwave background [5], revealing a severe deficit of the SM in explaining the observed matter abundance of the universe. This motivates searches of undiscovered sources of CP violation, spanning different sectors of particle physics today [6–8].

A non-zero value of the electric dipole moment (EDM) of a non-degenerate system, such as the neutron, would violate time-reversal (T) symmetry. This is equivalent to CP violation, assuming the CPT theorem [9] fundamental to relativistic quantum field theories. EDMs have been searched in different physical systems, such as the neutron, the electron and atoms [8], however no finite EDM has ever been observed. They have complementary sensitivities to undiscovered sources of CP violation, which would manifest differently in the different systems. Effective field theories provide a framework to interpret the experimental limits and

*e-mail: thiguchi@rcnp.osaka-u.ac.jp

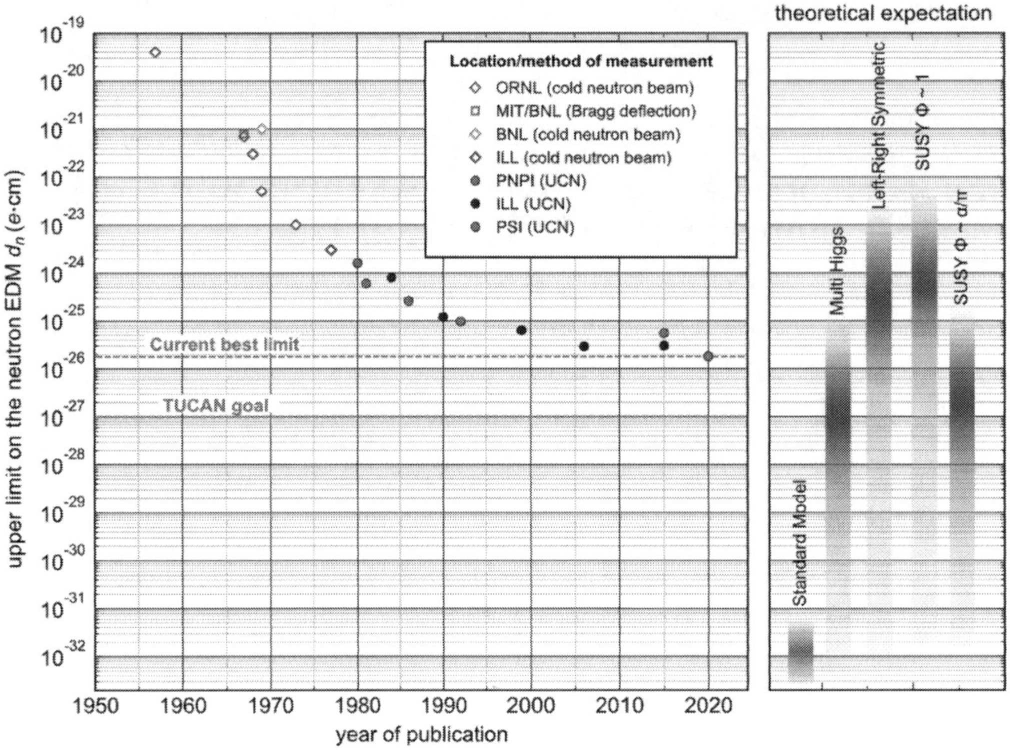

Figure 1. Historical development of the experimental upper limit on the nEDM. The method and the location of the experiments are indicated in the legend. The theoretical predictions according to the SM and BSM theories are shown on the right. The SM expectation is based on Ref. [8]. The expected values from the BSM models are based on Ref. [11].

place stringent constraints on super-symmetry models and other theories beyond the Standard Model [10].

In Fig. 1, development of the experimental upper limit on the neutron EDM (nEDM) over time is plotted together with theoretical expectations. The prediction based on SM CP violation is on the order of $10^{-32}e \cdot$ cm. On the other hand, the BSM theories expect orders of magnitude larger nEDM, which could be in reach of future experiments. The experimental method and the location of each measurement are indicated in the legend. Since the first direct measurement by Smith, Purcell and Ramsey [12] until a result obtained in 1977 [13], beams of cold neutrons were used. Since the 1970s, ultracold neutrons (UCNs) have been used for nEDM measurements, enabling the experimentalists to reach a precision on the order of $10^{-26}e \cdot$ cm. The UCNs refer to neutrons with kinetic energies $\lesssim$ 300 neV. Because of their extremely low energies, UCNs are totally reflected on appropriately-chosen material surfaces. As the consequence, they can be stored in a measurement volume for times on the order of 100 s, which greatly benefits the statistical sensitivity of the nEDM measurements. Such stored UCNs are also used at an experiment at PSI where the current best limit was obtained [14].

The major limiting factor of recent nEDM measurements is the number of UCNs available for the measurements, which is typically on the order of $N{\sim}10^4$ per cycle. Therefore, an intense UCN source is one of the keys for a further development in the field.

1.2 The TRIUMF Ultra-Cold Advanced Neutron collaboration

The international TUCAN (TRIUMF Ultra-Cold Advanced Neutron) collaboration was formed with the aims to build a high-intensity UCN source at TRIUMF and to measure the nEDM with $10^{-27} e \cdot$cm precision, thus to improve the current limit by an order of magnitude.

The principle of nEDM measurement is based on spin-precession frequency measurement under a weak ($\sim\mu$T) magnetic field $\boldsymbol{B}$ and a strong ($\sim$10 kV/cm) electric field $\boldsymbol{E}$. The interaction of the magnetic moment μ_n and the EDM d_n of the neutron with the external fields is represented by a Hamiltonian

$$H = -\boldsymbol{\mu}_n \cdot \boldsymbol{B} - \boldsymbol{d}_n \cdot \boldsymbol{E}. \tag{2}$$

To extract the value of d_n, the spin-precession frequencies are compared between configurations where the electric and the magnetic fields are parallel ($\uparrow\uparrow$) and anti-parallel ($\uparrow\downarrow$) to each other. Denoting these frequencies as $\omega_{\uparrow\uparrow}$ and $\omega_{\uparrow\downarrow}$, respectively, d_n can be expressed as

$$d_n = \frac{\hbar(\omega_{\uparrow\uparrow} - \omega_{\uparrow\downarrow})}{4|\boldsymbol{E}|}. \tag{3}$$

The spin precession frequencies $\omega_{\uparrow\uparrow}$ and $\omega_{\uparrow\downarrow}$ are measured by the Ramsey's technique of separately oscillating fields [15] where two $\pi/2$ pulses separated by a free precession time are used to obtain a spin resonance. In this method, the statistical uncertainty of the nEDM δd_n is determined by the free-precession time T, the number of neutrons N as

$$\delta d_n \sim \frac{\hbar}{|\boldsymbol{E}|T\sqrt{N}}. \tag{4}$$

Our high-intensity UCN source is expected to provide $N \sim 10^6$ UCNs per cycle to the measurement cell. Together with the other parameters based on realistic assumptions, $|\boldsymbol{E}| \sim 10$ kV/cm and $T \sim 100$ s, $\delta d_n \sim 10^{-25} e \cdot$ cm per cycle is obtained. The statistical sensitivity goal of $10^{-27} e \cdot$ cm will be reached in about 400 days of measurement, corresponding to $\sim 10^4$ cycles.

The UCN production scheme of TUCAN was demonstrated with a prototype UCN source operated at TRIUMF in 2017–2019 [18]. Currently, we are building an upgraded UCN source, and in parallel, developing subsystems of the nEDM spectrometer. In the following sections, the UCN production scheme of TUCAN is explained, and some of recent activities towards the planned nEDM measurement with the new UCN source are presented.

2 Provision of Ultracold Neutrons

2.1 Principles of UCN production

The UCN production scheme of TUCAN [19] depicted in Fig. 2 is a combination of spallation neutron production and the super-thermal method using superfluid helium (He-II) [20]. First, a proton beam from an accelerator impinges on a heavy metal target to produce fast neutrons with MeV kinetic energies. They are thermally cooled by layers of neutron moderators to meV energies. The resultant cold neutrons are converted to UCNs via inelastic down-scattering in He-II [20]. The dispersion relation of He-II, as shown in Fig. 3, is characterized by photon and roton excitations. The He-II dispersion curve intersects with that of the free neutron at $\approx$ 12 K, corresponding to 1 meV. If a neutron with the corresponding momentum enters He-II, it can lose most of its energy by creating one phonon, and be converted to a UCN. After becoming a UCN, it can be guided to a downstream experimental area.

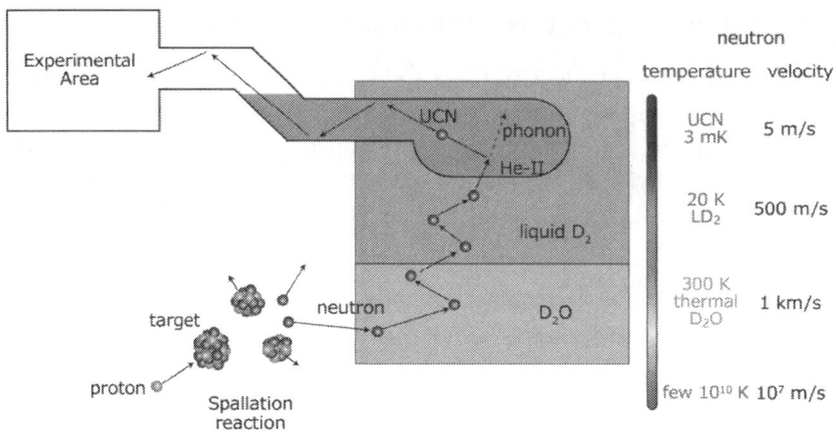

Figure 2. The UCN production scheme of TUCAN. Fast neutrons with MeV energies (corresponding to $\sim 10^{10}$ K temperature) are produced and cooled by a layer of neutron moderators to meV energies ($\sim$10 K). In a bottle containing He-II, cold neutrons are converted to UCNs and then extracted to the downstream experimental area by reflections inside a UCN guide. Reproduced from Ref. [22].

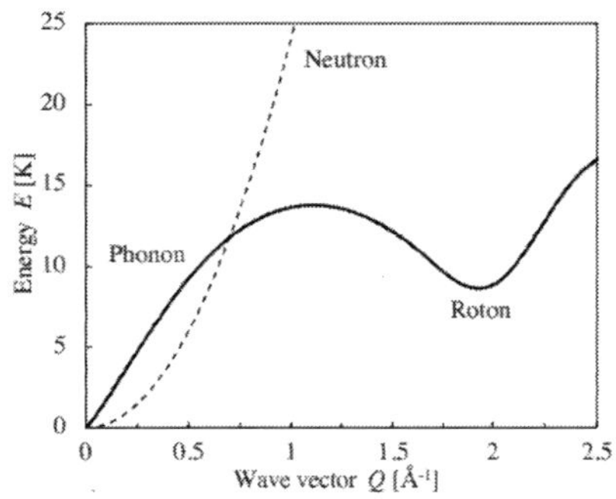

Figure 3. The dispersion curve of phonons and rotons excited in He-II (solid line). It intersects the free neutron curve (dashed line) at a temperature around 12 K. Cold neutrons around this energy are most efficiently converted to UCNs. Adapted from Ref. [16], based on data from Ref. [17].

One of the crucial parameters in this process is the He-II temperature $T_{\text{He-II}}$, because the major UCN loss process is due to heating by up-scattering with phonons, whose rate τ_{up}^{-1} scales as $\tau_{\text{up}}^{-1} \propto T_{\text{He-II}}^7$. For efficient UCN production, $T_{\text{He-II}} \lesssim 1$ K needs to be satisfied. Another important factor is a flux of cold neutrons around the critical energy of 1 meV.

2.2 Achievements with the prototype UCN source

The UCN production method described in the last section was tested with the prototype UCN source, which was originally developed at RCNP, Osaka University and relocated to TRIUMF in 2016. With this prototype source, we succeeded in producing the first UCNs at TRIUMF in 2017 [18]. Results of this beamtime are presented in Fig. 4. Fig. 4a shows typical detector counts. After beam irradiation of the target for 60 s, UCNs stored in the production volume were extracted towards a downstream detector via about 6 m long UCN guides. Shown in Fig. 4b is a relation between the UCN yield and the proton beam current. The UCN yield scales proportionally to the beam current up to 1 μA. For higher beam currents, however, it drops from this scaling. Under these conditions, temperature increases in the He-II converter were observed. Therefore, the reduction of the UCN yield is due to a heat load on the He-II converter beyond the cooling capacity of the cryostat, which increased the He-II temperature and consequently the up-scattering UCN loss. The prototype source was thus limited by the

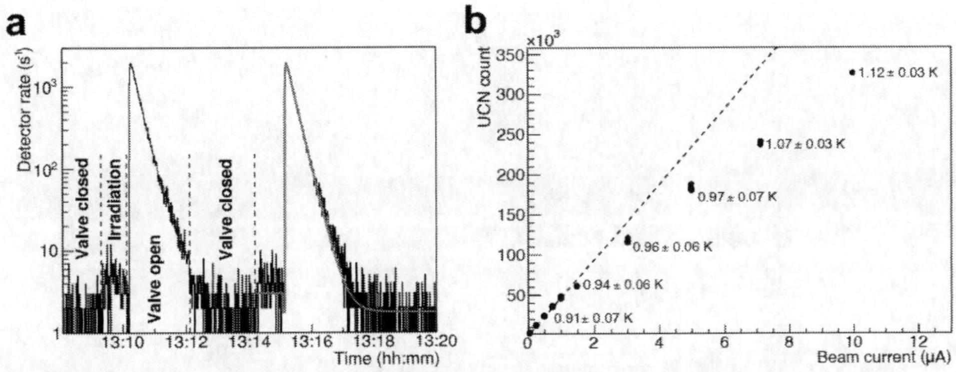

Figure 4. Results obtained during the beamtime 2017 by the prototype UCN source. Reproduced from Ref. [18]. **a.** Detector rate in typical two cycles during operation with a 1 μA beam current. After proton beam irradiation for 60 s, the valve towards the detector is opened for 120 s. While the valve is open, UCNs stored in the irradiation volume are counted by the detector. **b.** The UCN counts extracted from the source after irradiating the spallation target for 60 s with different beam currents. The labels indicate the peak helium temperatures reached under each irradiation condition.

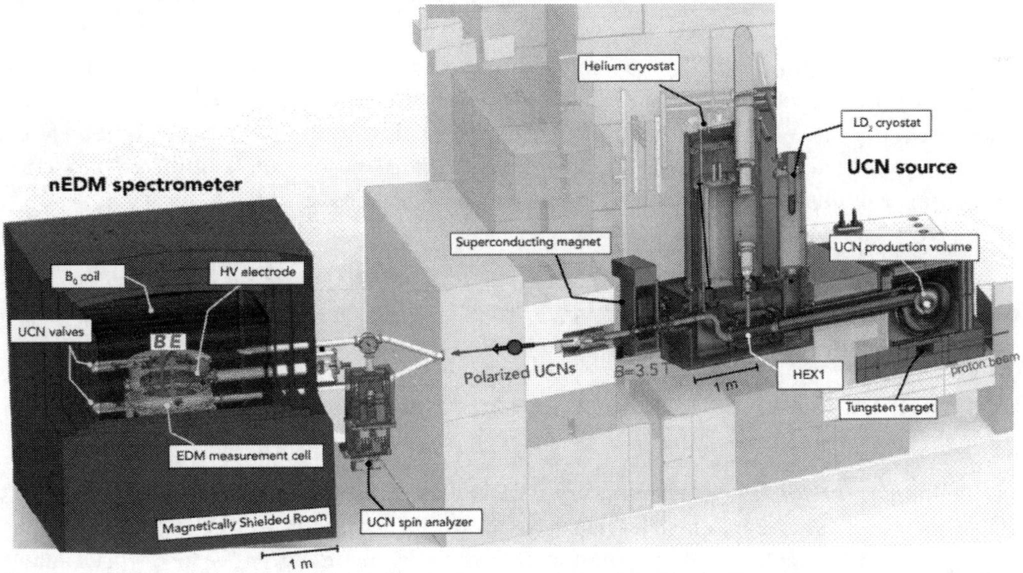

Figure 5. Overview of the new UCN source and the nEDM spectrometer of the TUCAN collaboration. Spallation neutrons from the tungsten target are moderated to cold neutron energies and then converted to UCNs in the production volume filled with He-II. The UCNs are then polarized by a 3.5 T magnetic field of the superconducting magnet and led into the nEDM spectrometer through UCN guides.

cooling power of its cryostat. A typical UCN yield achieved with nominal beam parameters of 483 MeV and 1 μA was 47 500 UCN/cycle for a batch-mode operation with 60 s irradiation, and 1 500 UCN/s in a continuous extraction mode.

2.3 Development of the new UCN source

Based on the experiences with the prototype source, the new UCN source is being developed (Fig. 5), which addresses the limitation of the prototype. The major upgrade is on the helium cryostat [22], which now enables UCN production at 40 μA beam current and increases the

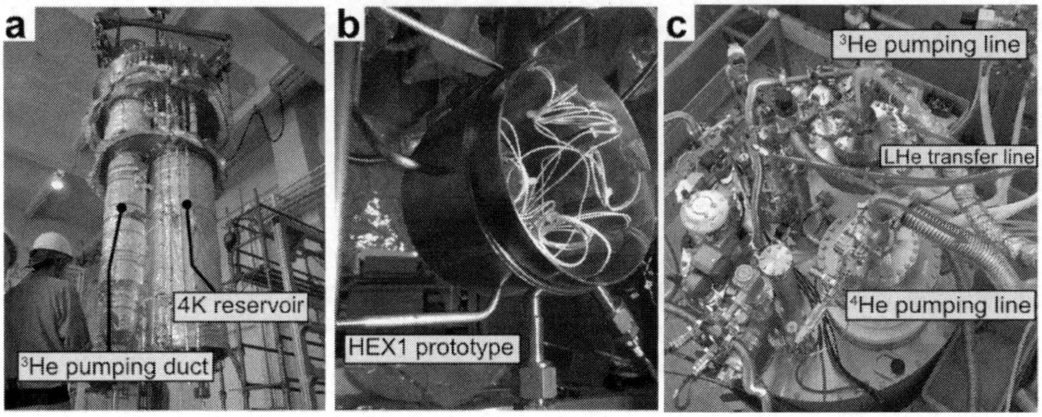

Figure 6. Photographs during the installation of the helium cryostat at KEK for cryogenic tests. For more details of the structure of the cryostat, see Ref. [22]. **a.** The cryostat while the heat shields are being assembled. **b.** The prototype HEX1 attached at the bottom of the cryostat. It has a 1/10 length of the final HEX1 shown in Fig. 5. **c.** The top flange of the cryostat after being installed in a test pit.

UCN yield by a factor of 40. Replacement of the cold-neutron moderator from solid heavy water (sD_2O) to liquid deuterium (LD_2) adds another factor of 3 gain in the UCN yield, because LD_2 produces a higher flux of cold neutrons at 1 meV than sD_2O. Together with the other gain factors, such as an increased production volume and optimized moderator geometries, a production rate of 1.4 to 1.6×10^7UCN/s is expected [21].

3 Highlights of recent progress

3.1 Construction and tests of the helium cryostat for the new UCN source

As mentioned in Sec. 2.3, the helium cryostat is one of the crucial ingredients of the new UCN source. A highly-efficient ^{3}He-^{4}He cryostat with a cooling power of 10 W at $\sim$1 K was developed, which enables UCN production with a 40 μA beam current. The cryostat mainly consists of seven types of heat-exchangers. Its central item is a copper heat exchanger named HEX1, which interfaces between a 0.8 K ^{3}He pot and the UCN production volume filled with He-II. The heat load is removed by latent heat loss due to nucleate boiling of ^{3}He in HEX1. The other heat exchangers are designed to efficiently recover the enthalpy of outgoing fluid at each temperature stage [22, 23]. The construction of the cryostat components commenced in 2019. As they were built, performance of each heat exchanger was characterized by cryogenic tests to validate thermal calculations and simulations used for the design [23]. From September 2020 to March 2021, the full cryostat assembly was tested with natural helium at KEK (Fig. 6). It showed excellent performance in pre-cooling efficiency and its low static heat load. The lowest temperature achieved with ^{4}He was 1.23 K, corresponding to 0.65 K for ^{3}He at the same saturation vapour pressure. In total four cool downs were performed to characterize boiling heat transfer of ^{4}He with different geometries and supply configurations of HEX1 prototypes. Based on these results, the final model of HEX1 is under development. The cryostat was transported to TRIUMF in August 2021. It will be tested with ^{3}He and high-capacity pumps prior to the commissioning with the other parts of the UCN source.

3.2 Magnetic field measurements and simulations for the installation of the magnetically shielded room

One of the most crucial systematic effects which should be controlled for the nEDM measurement is the magnetic field. In deriving Eq. (3), the magnetic fields between the two field configurations ($\uparrow\uparrow$) and ($\uparrow\downarrow$) are assumed to be identical. However, in the actual measurement, variations and inhomogeneities of the magnetic field impose systematic limitations on the measurement. Regarding the magnetic-field stability, the cycle-to-cycle magnetic field stability on the order of 10 fT/cycle is required for an nEDM measurement with the target sensitivity. This will be achieved by a high-performance magnetic shield and correction of magnetic-field variations by the use of a ^{199}Hg co-magnetometer [24]. The magnetic-field fluctuations inside the magnetic shield are required to be on the order of 1 pT/cycle. To meet this requirement, a magnetically shielded room (MSR) consisting of four layers of mumetal has been designed (Fig. 5). The design shielding factor on the order of 10^5 is validated by finite-element-analysis (FEA) simulations. In addition, the residual magnetic field and its gradient in the central $1 \times 1 \times 1$ m^3 volume will be < 1 nT and 100 pT/m, respectively.

Although such magnetic shields have been developed and tested at other facilities [25, 26], a unique challenge in case of TUCAN results from a strong background field up to $|\boldsymbol{B}| \approx 370\,\mu$T from the TRIUMF cyclotron. This induces a magnetic flux in the outermost layer of the MSR on a level close to the saturation induction of mumetal, which is about 700 mT. To characterize the spatial distribution of the background magnetic field, three-dimensional magnetic field mapping was performed around the planned installation location of the MSR in August 2019 (when the cyclotron was operational) and February 2020 (during cyclotron shutdown) (Fig. 7a). Fig. 7b and c show the spatial distributions of the vertical component of the background magnetic field when the cyclotron was operated. The main part of the background field is approximated by a dipole field originating from the cyclotron. Based on this data, a set of compensation coils are designed, which will reduce the magnetic flux in the mumetal layer and ensure the performance of the MSR in the presence of the background field. Results of FEA simulations shown in Fig. 8 demonstrate that sufficient compensation can be achieved by a simple combination of square coils. The MSR and the compensation coil system will be installed by November 2022. Thereafter, installation and tests of inner coils and magnetometers in the MSR will begin.

3.3 Component tests with cold and ultracold neutrons at J-PARC/MLF

Some components of the new UCN source and the nEDM spectrometer have been developed by the use of cold and ultracold neutrons available at the Materials and Life Science Facility (MLF) of J-PARC. The beamline BL05 [27] of the MLF is equipped with a Doppler-shifter-type UCN source, which produces pulsed UCNs by decelerating very cold neutrons (kinetic energies of $\sim$100 μeV) with a rotating neutron mirror moving backwards in respect to the incident neutrons [28].

In June 2020, a systematic measurement was conducted to characterize UCN transmission through cylindrical guides coated with Nickel phosphorus (NiP). The transmission rate of UCNs through a guide was measured for different incident angles. By comparing the measured transmission to Monte-Carlo simulations, a realistic model of diffuse reflection of UCNs due to surface roughness of the guide was developed [29].

The pulsed UCNs from the Doppler shifter can also be used to test storage lifetime of critical inner components. A pilot experiment was successfully performed in February 2021, where a one-meter UCN guide was used as a storage volume [29]. A storage experiment of a prototype of the EDM measurement cell (Fig. 5) is planned in spring 2022.

https://doi.org/10.1051/epjconf/202226201015

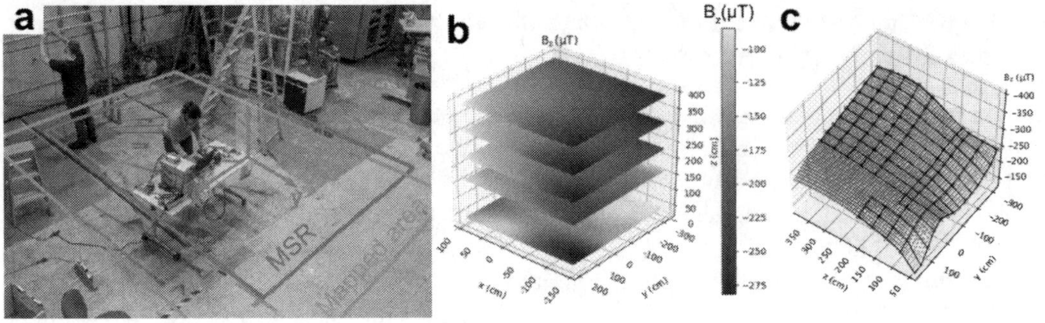

Figure 7. a. Magnetic field mapping campaign performed at the planned location of the TUCAN MSR in the TRIUMF Meson Hall. The mapped area, the planned location of the MSR, and the coordinate system are indicated in the photograph. **b.** The magnetic field distribution obtained by the measurement when the cyclotron was operational. The three-dimensional distribution of the vertical component B_z is displayed. **c.** A two-dimensional cut of the data of b. at a plane $x = 182$ cm. The data points connected with solid black lines are those obtained by actual measurements. Data indicated by the coloured dots are obtained by interpolating the measured data points. Some local field distributions are observed near the floor height $z = 0$. The field distribution at high z positions is found to be approximated by a dipole field originated from the TRIUMF cyclotron, which is located in $-x$ direction in this coordinate system.

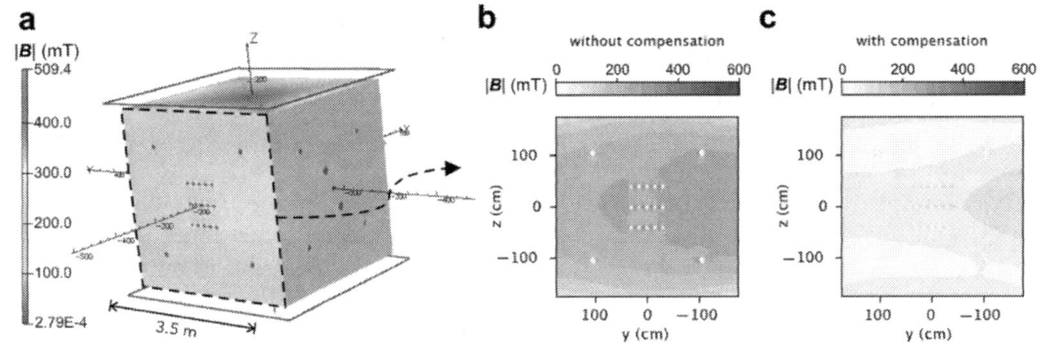

Figure 8. Results of finite-element magnetic field simulations for design of the compensation coil system surrounding the MSR. **a.** The MSR and the coils are placed in a background magnetic field reproduced from an actual magnetic field map. The scale indicates the norm of the in-plane magnetic flux density $|\boldsymbol{B}|$ of the outermost layer of the mu-metal sheet when the coils are not charged. **b.** The $|\boldsymbol{B}|$ distribution in one of the planes when the coils are not charged. Around the holes of the mumetal plates, it reaches up to $|\boldsymbol{B}| \approx 500$ mT. **c.** The $|\boldsymbol{B}|$ distribution in the same plane when the upper and lower coils are charged to 1000 Ampere-turns. The in-plane flux density $|\boldsymbol{B}|$ is reduced to about 100 mT or less.

Another item under development is the UCN spin analyzer, whose main component is a thin iron film magnetized by a surrounding magnet. When a UCN traverses the film, it experiences a spin-dependent potential expressed as

$$V_{\uparrow,\downarrow} = V_{\mathrm{F}} \pm \mu_n \cdot B \approx 209 \mp 60 \text{ neV/T} \cdot B, \tag{5}$$

where $\mu_n \approx -60$ neV/T and $V_{\mathrm{F}} \approx 209$ neV represent the magnetic moment of the neutron and the Fermi potential of iron, respectively. Because kinetic energies of UCNs are $\lesssim 300$ neV, a magnetic flux density $B \approx 2$ T produced in a fully saturated iron film makes it a filter which selectively transmits one of the spin components. Iron thin films used for this purpose are required to be saturated with a small (~ 10 mT) magnetic field, and have a saturation induction close to 2 T. The ion-beam sputtering (IBS) facility at the Institute for Integrated Radiation

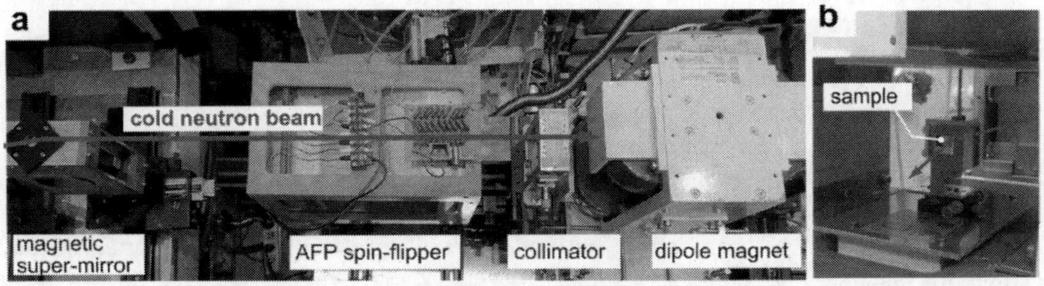

Figure 9. Setup of the cold-neutron reflectometry measurement of iron thin film samples. **a.** The main components of the setup. The cold neutron beam is polarized by a magnetic super-mirror and goes through an AFP spin flipper and a collimator and then injected on a sample placed in a dipole magnet. The neutrons are detected by a two-dimensional detector based on resistor type photomultiplier tubes (RPMT) placed 67 cm downstream of the dipole magnet (not shown in the picture). **b.** View of the dipole magnet from the downstream side. A 20×30 mm^2 sample of silicon substrate iron film is installed in a holder and placed in the magnet.

and Nuclear Science, Kyoto University (KURNS) [30] was used to produce magnetically soft iron thin films on silicon or aluminum substrates. In July 2021, samples of silicon-substrate iron thin films were characterized by a cold-neutron reflectometry measurement on the low-divergence beam branch of BL05 [27]. In this method, neutrons are injected on a surface of a sample with a small incident angle to obtain the reflectivity as a function of the wavevector transfer $R(q)$. The Fermi potential of the sample is obtained from the critical wavevector transfer q_c, below which the neutrons are totally reflected. In this experiment, the incident neutrons are polarized by a magnetic super-mirror [31] and injected on a sample placed in a magnetic field of a dipole magnet (Fig. 9). A spin flipper based on the principle of adiabatic free passage [32] is placed between the super-mirror and the dipole magnet to flip the neutron spin when activated. By measuring the reflectivity for each spin component under different magnetic fields, the effective potential $V_{\uparrow,\downarrow}$ of Eq. (5) is extracted. Samples with different iron-layer thicknesses (30, 50, 90 nm) are measured under a magnetic field of 0–8 mT. From the result, the effective potential $V_\downarrow \approx 328$ neV (corresponding to ≈ 2.0 T magnetic flux density) is obtained with a few mT of the applied field. As the next step, a polarization experiment of pulsed UCNs provided by the Doppler shifter is planned in spring 2022.

4 Conclusions and outlook

The measurement of the nEDM provides a direct test of time-reversal violation and provides constraints on the BSM theories. The TUCAN collaboration aims to measure the nEDM with $10^{-27} e \cdot$ cm precision to provide an order-of-magnitude improvement upon the current best experimental limit. To overcome the statistical limitation of the current nEDM measurement with stored UCNs, TUCAN is developing a new high-intensity UCN source based on spallation neutron production and the super-thermal UCN conversion with He-II.

There has been major progress on the core components of the new UCN source and the nEDM spectrometer. Among them, the status of the helium cryostat, magnetic-field subsystems including the MSR, and component tests using at J-PARC/MLF have been reported.

Currently, we aim to install the major part of the new UCN source on the TRIUMF beamline by October 2022, and to start commissioning the nEDM spectrometer in 2023.

Acknowledgements

The author thanks C. Marshall, T. Hessels, S. Horn and D. Rompen (TRIUMF) for their engineering support, M. Hino (KURNS) for iron thin film production and valued support on the cold-neutron reflectometry measurement.

The neutron experiments at the Materials and Life Science Experimental Facility of J-PARC were performed under S-type project of KEK (Proposal No. 2019S03). The iron thin film fabrication work has been carried out under the visiting Researcher's Program of the Institute for Integrated Radiation and Nuclear Science, Kyoto University.

The work of the TUCAN collaboration is supported by the Canada Foundation for Innovation (CFI), the Canada Research Chairs program, the Natural Sciences and Engineering Research Council of Canada (NSERC), B.C. Knowledge Development Fund (BCKDF), Research Manitoba, the Japan Society for the Promotion of Science (JSPS), Yamada Science Foundation, and the RCNP Collaboration Research network (COREnet) program.

References

[1] T. D. Lee and C. N. Yang, Phys. Rev. **104**, 254 (1956); Phys. Rev. **106**, 1371 (1957).
[2] C. S. Wu et al., Phys. Rev. **105**, 1413 (1957).
[3] A. D. Sakharov, Pisma Zh. Eksp. Theor. Fiz. **5**, 32 (1967).
[4] L. Canetti, M. Drewes and M. Shaposhnikov, New. J. Phys. **14**, 095012 (2012).
[5] P. A. R. Ade et al. [Planck Collaboration], Astron. Astrophys. **594**, A13 (2016).
[6] R. Aaij et al. [LHCb Collaboration], Phys. Rev. Lett. **124**, 03801 (2020).
[7] The T2K Collaboration, Nature **580**, 339 (2020).
[8] T. E. Chupp et al., Rev. Mod. Phys. **91**, 015001 (2019).
[9] G. Lüders, Ann. Phys. **2**, 1, (1957).
[10] J. Engel, M. J. Ramsey-Musolf and U. van Kolck, Prog. Part. Nucl. Phys. **71**, 21 (2013).
[11] J. M. Pendlebury and E. A. Hinds, Nucl. Inst. Meth. A, **440**, 471 (2000).
[12] J. H. Smith, E. M. Purcell, and N. F. Ramsey, Phys. Rev. **108**, 120 (1957).
[13] W. B. Dress et al., Phys. Rev. D **15**, 9 (1977).
[14] C. Abel et al., Phys. Rev. Lett. **124**, 081803 (2020).
[15] N. F. Ramsey, Phys. Rev. **78**, 69 (1950).
[16] S. Kawasaki, T. Kikawa and R. Matsumiya, High Energy News (KEK) **37**, 11 (2018).
[17] R. J. Donnelly and C. F. Barenghi, J. Phys. Chem. Ref. Data **27**, 1217 (1998).
[18] S. Ahmed et al. [TUCAN Collaboration], Phys. Rev. C **99**, 025503 (2019).
[19] Y. Masuda et al., Phys. Rev. Lett., **108**, 134801 (2012).
[20] R. Golub and J. M. Pendlebury, Phys. Lett. **62A**, 337 (1977).
[21] W. Schreyer et al., Nucl. Inst. Meth. A **959**, 163525 (2020).
[22] S. Kawasaki et al., IOP Conf. Ser.: Mater. Sci. Eng. **755**, 012140 (2020).
[23] T. Okamura et al., IOP Conf. Ser.: Mater. Sci. Eng. **755**, 012141 (2020).
[24] K. Green et al., Nucl. Inst. Meth. A **404**, 381 (1998).
[25] I. Altarev et al., J. App. Phys. **117**, 183903 (2015).
[26] S. Knappe-Grueneberg et al., J. App. Phys. **103**, 07E925 (2008).
[27] K. Mishima et al., Nucl. Inst. Meth. A **600**, 342 (2009).
[28] S. Imajo et al., Prog. Theor. Exp. Phys. **2016**, 013C22 (2016).
[29] T. Higuchi, S. Imajo et al., Annual report of Research Center for Nuclear Physics, Osaka University 2020 (2021).
[30] M. Hino et al., Nucl. Inst. Meth. A **797**, 265 (2015).

[31] M. Hino et al., Physica B **385-386**, 1187 (2006).
[32] S. V. Grigoriev, A. I. Okorokov and V. V. Runov, Nucl. Inst. Meth. A **384**, 451 (1997).

EPJ Web of Conferences **262**, 01016 (2022) https://doi.org/10.1051/epjconf/202226201016
EXA2021

Main Features of the SIDDHARTA-2 Apparatus for Kaonic Deuterium X-Ray Measurements

Marlene Tüchler[1,*], *Florin* Sirghi[2,3,**], *M.* Bazzi[2], *D.* Bosnar[4], *M.* Bragadireanu[3], *M.* Carminati[5], *M.* Cargnelli[1], *A.* Clozza[2], *C.* Curceanu[2], *G.* Deda[5], *R.* Del Grande[2,6,7], *L.* De Paolis[2], *C.* Fiorini[5], *C.* Guaraldo[2], *M.* Iliescu[2], *M.* Iwasaki[8], *P.* King[5], *P.* Levi Sandri[2], *J.* Marton[1], *M.* Miliucci[2], *P.* Moskal[9], *F.* Napolitano[2], *S.* Niedzwiecki[9], *K.* Piscicchia[7,2], *A.* Scordo[2], *F.* Sgaramella[2,10], *H.* Shi[1], *M.* Silarski[9], *D.* Sirghi[2,3], *M.* Skurzok[9,2], *A.* Spallone[2], *O.* Vazquez Doce[2,6], *E.* Widmann[1], and *J.* Zmeskal[1]

[1]Stefan Meyer Institute for Subatomic Physics, Vienna, Austria
[2]INFN, Laboratori Nazionali di Frascati, Frascati (Roma), Italy
[3]Horia Hulubei National Inst. of Physics and Nuclear Engineering, Bucharest, Romania
[4]Department of Physics, Faculty of Science, University of Zagreb, Zagreb, Croatia
[5]Politecnico di Milano, Dipartimento di Elettronica, Informazione e Bioingegneria and INFN Sezione di Milano, Milano, Italy
[6]Excellence Cluster Universe, Technische Universiät München, Garching, Germany
[7]Centro Ricerche Enrico Fermi - Museo Storico della Fisica e Centro Studi e Ricerche "Enrico Fermi", Roma, Italy
[8]RIKEN, Tokyo, Japan
[9]M. Smoluchowski Institute of Physics, Jagiellonian University, Kraków, Poland
[10]Università degli Studi di Roma, Tor Vergata, Rome, Italy

Abstract. The low-energy, non-perturbative regime of QCD can be studied directly by X-ray spectroscopy of light kaonic atoms. The SIDDHARTA-2 experiment, located at the DAΦNE collider, aims to measure the $2p \rightarrow 1s$ transition in kaonic deuterium for the first time to extract the antikaon-nucleon scattering lengths. This measurement is impeded, inter alia, by the low K^-d X-ray yield. Hence, several updates have been implemented on the apparatus to increase the signal-to-background ratio, which are discussed in detail in this paper: a lightweight gas target cell, novel Silicon Drift Detectors for the X-ray detection with excellent performance, and a veto system for active background suppression. The experiment has undergone a first preparatory run during DAΦNE's commissioning phase in 2021, concluding with a successful kaonic helium measurement.

1 Scientific Motivation

Light kaonic atoms provide a unique key to study the antikaon-nucleon interaction at threshold and hence to access the low-energy, non-perturbative regime of the strong interaction at zero relative energy. In kaonic hydrogen and deuterium systems, the kaon interacts via the strong interaction with the nucleons, which induces a shift in energy as well as a broadening

*e-mail: marlene.tuechler@oeaw.ac.at
**e-mail: Sirghi.FlorinCatalin@lnf.infn.it

of the kaonic atom ground state. Through measurements of the K_α transition in these light kaonic atoms, the $1s$ shift ϵ_{1s} and width Γ_{1s} can be observed. A combination of the results for kaonic hydrogen and kaonic deuterium allows to extract the isospin-dependent antikaon-nucleon scattering lengths a_0 and a_1, which provide crucial constraints for the prevalent theoretical models. An overview of the experimental work on kaonic atoms is given in [1].

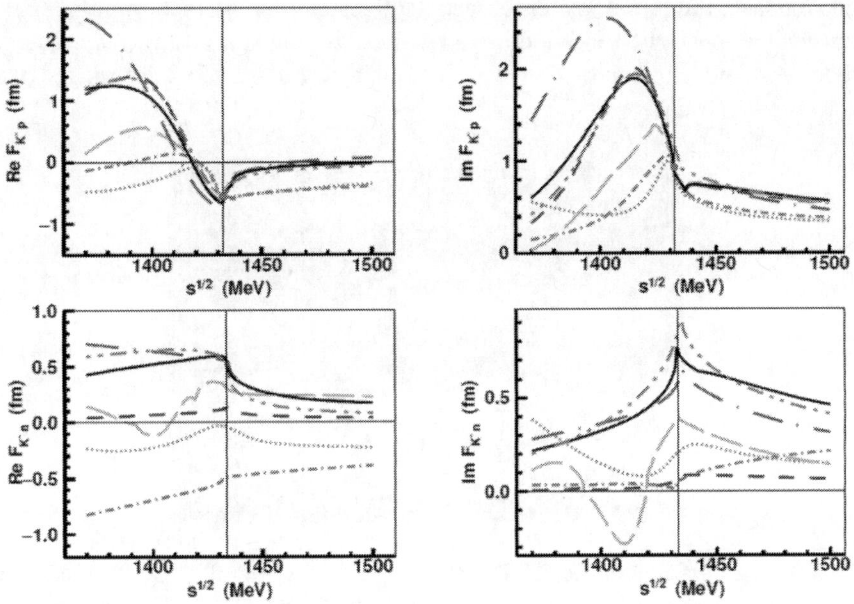

Figure 1. The elastic scattering amplitudes for the K^-p (*top*) and K^-n (*bottom*) cases given by different chiral models: Bonn (dotted and dot-dashed) [4], Murcia (dashed and long-dashed) [5], Prague (dot-long-dashed) [3], Barcelona (dot-dot-dashed) [6], and Kyoto-Munich (continuous) [2]. *From [7].*

Fig. 1 shows calculations from different chiral effective models [2–6] for the elastic scattering amplitudes for the proton and neutron case, respectively [7]. For the antikaon-proton calculations, the successful kaonic hydrogen measurement performed by the SIDDHARTA collaboration in 2009 [8] provided a significant constraint on the parameters of the models describing the real and imaginary part of the K^-p elastic scattering amplitudes to reproduce the SIDDHARTA data points (Fig. 1 top panels). Therefore, the SIDDHARTA-2 collaboration now aims to provide the same constraint for the K^-n case by measuring the $2p \rightarrow 1s$ transition in kaonic deuterium.

The SIDDHARTA-2 experiment is located at the DAΦNE collider at Laboratori Nazionali di Frascati (LNF) in Italy. The challenge of the K^-d measurement is the low expected kaonic deuterium X-ray yield of $Y_{K_\alpha} < 0.0039\%$ [9], as well as the broad ground state width as compared to kaonic hydrogen. Therefore, the experimental apparatus has undergone several updates to improve the signal-to-background ratio by at least one order of magnitude, which is crucial for the K^-d measurement. These updates include:

- a luminosity monitor to study the beam and background conditions,
- a lightweight, cryogenic gaseous target cell,
- 48 arrays of newly developed monolithic Silicon Drift Detectors (SDDs) for the detection of X-rays,
- and a dedicated veto system.

2 The SIDDHARTA-2 Apparatus

The SIDDHARTA-2 apparatus was installed at the DAΦNE collider in April 2019. A schematic drawing of the apparatus is shown in Fig. 2 including all the main components. The target cell is shown in yellow, surrounded by the X-ray detectors and inner layer of the veto system. Outside of the vacuum chamber, the outer layer of veto detectors is shown. A luminosity monitor, the luminometer, is installed close to the interaction point [10]. Due to the special feature of DAΦNE producing the Φ mesons almost at rest, the charged kaons are emitted back-to-back, and can be detected in coincidence using the kaon trigger to suppress asynchronous background.

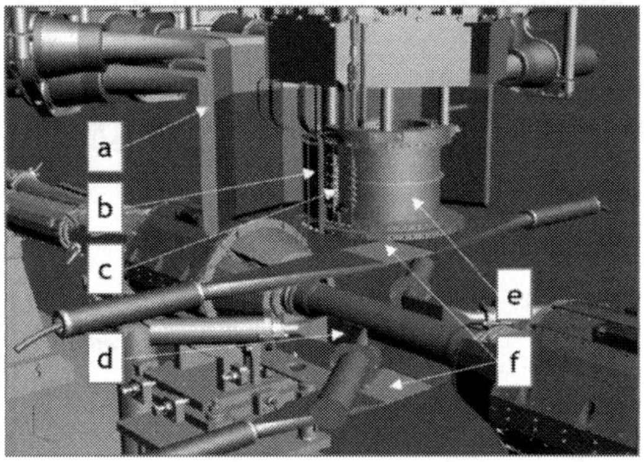

Figure 2. Schematic drawing of the SIDDHARTA-2 apparatus with the veto systems (a, b), X-ray detectors (c), luminosity monitor (d), target cell (e) and kaon trigger (f).

2.1 The Target Cell

Since X-ray losses due to the Stark effect severely aggravate the measurements in light kaonic atoms, the target gas density has to be optimised to strike a balance between a sufficiently high kaon stopping power and minimal X-ray losses. Therefore, the working parameters of the SIDDHARTA-2 target cell are a temperature of 30 K and a deuterium density of 1.5% liquid density (LDD). For a 90% transmission of 8 keV X-rays to the X-ray detectors, the sidewalls are made from two layers of 50 μm Kapton foils glued together with epoxy adhesive, resulting in a total thickness of $\sim$ 150 μm. The target cell with a height of 120 mm and a diameter of 144 mm is shown in Fig. 3.

Figure 3. The lightweight SIDDHARTA-2 target cell with Kapton sidewalls and the mounting structures for the SDDs.

2.2 The Silicon Drift Detectors

The X-ray detection system of SIDDHARTA-2 consists of 48 arrays of newly developed SDDs. Each array features eight cells of 8×8 mm^2 (Fig. 4 left panel), resulting in a total of 384 read-out channels and a total active area of 246 cm^2. In comparison to the SDDs used for the kaonic hydrogen measurement in SIDDHARTA, the new detectors have been structurally updated. Before, an n-channel JFET was implemented directly on the anode structure of the SDD. For SIDDHARTA-2, however, a preamplifier based on a MOSFET technology, called CUBE, is implemented on the ceramic carrier structure of the SDD close to the anode (Fig. 4 *right*). This update allows for a lower operational temperature of the SDDs and a more stable operation at high rates, as well as lower drift times ($\sim$ 400 ns) and an improved energy resolution. Thanks to the geometry of the ceramic carrier, a solid angle of $\sim 2\pi$ is achieved.

Figure 4. *Left:* An SDD array with eight read-out channels. The geometry of the ceramic holding structure enables a compact arrangement of detectors around the target cell. *Right*: Schematic of an SDD chip with the CUBE (a) mounted on the ceramics (c), directly connected to the anode (b).

The SDDs were characterized at LNF and Stefan Meyer Institute, Vienna, in terms of their stability, linearity, timing performance and energy resolution [11–13]. Calibration measurements were performed, using an Fe-55 source activating a Ti foil. A typical calibration spectrum is shown in Fig. 5, were the characteristic X-ray lines for Ti (K_α peak at 4.5 keV) and Mn (K_α peak at 5.9 keV) are clearly visible. The Ca K_α peak at 3.7 keV (yellow) originates from the surrounding setup materials. The bottom panel of Fig. 5 shows the fit residuals. For 6 keV X-rays, an energy resolution of approximately 145 eV was measured at a temperature of 120 K.

2.3 The Veto Systems

The SIDDHARTA-2 experiment is equipped with a two-stage veto system to actively suppress the synchronous background. The purpose of the outer layer of the veto system, the Veto-1, is the distinction between signals originating from kaon stops in the target gas, and kaon stops in the surrounding materials. Since the stop of a kaon in a solid is much faster than in gas, timing information can be used to make this distinction. Therefore, a time resolution of the time delay with respect to the kaon trigger smaller than 1 ns (FWHM of the peak) has to be achieved. The Veto-1 system consists of twelve units of $(260 \times 110 \times 10)$ mm^3 plastic scintillators read out by photomultipliers (PMTs). To achieve the required time resolution despite the limited space around the vacuum chamber, mirrors and light guides are used to read out the scintillators on both ends (Fig. 6). In this way, a time resolution of (746 ± 53) ps (FWHM) is achieved [14].

The inner layer of the veto system, the Veto-2, is mounted directly behind the SDDs and is dedicated to the suppression of signals originating from minimum ionizing particles (MIPs) produced in the final kaon absorption on the nucleons. A background event in the X-ray region of interest can be produced should the MIPs traverse the SDDs on the edge of their

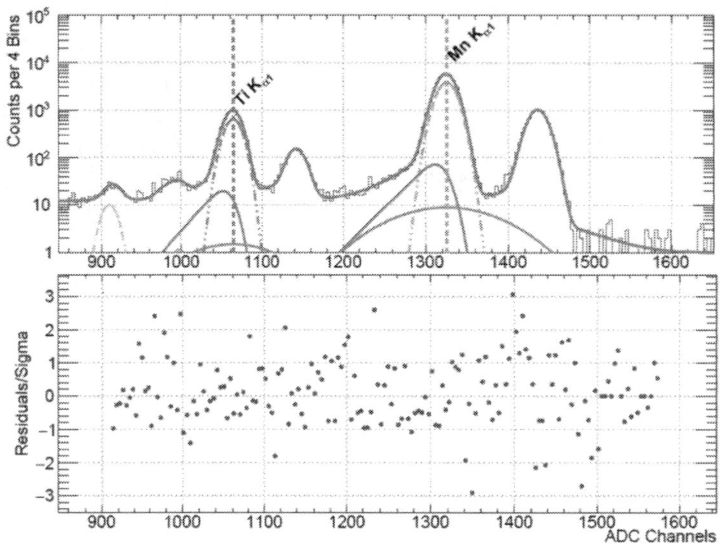

Figure 5. *Top:* Fitted SDD calibration spectrum obtained with an Fe-55 source and Ti foil. *Bottom:* Residuals from the fit of the calibration spectrum.

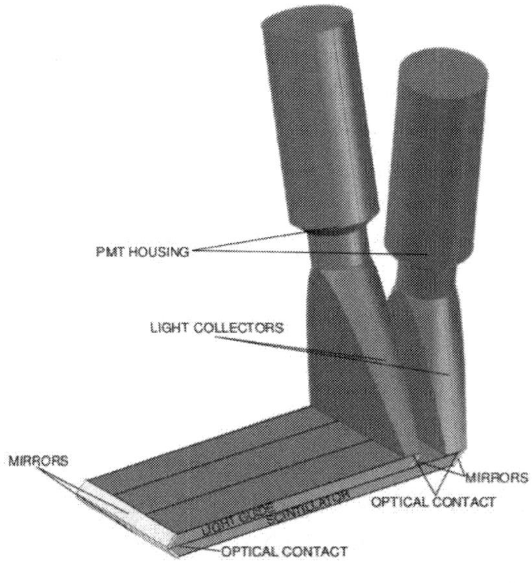

Figure 6. Detector unit of the Veto-1 system: A plastic scintillator, read out on both ends with PMTs using light guides. *From [14].*

active area. The spatial correlation between hits in the X-ray detectors and the Veto-2 system enables the rejection of these events [15]. The Veto-2 system consists of 24 detector units, each composed of four $(50 \times 12 \times 5)$ mm^3 plastic scintillators with Silicon Photomultiplier (SiPM) read-out. One of these units is shown in Fig. 7. Additionally, two pulsed LEDs are implemented per unit, allowing for an in-situ calibration of the Veto-2 detectors and for monitoring their performance, e.g. in terms of possible radiation damage.

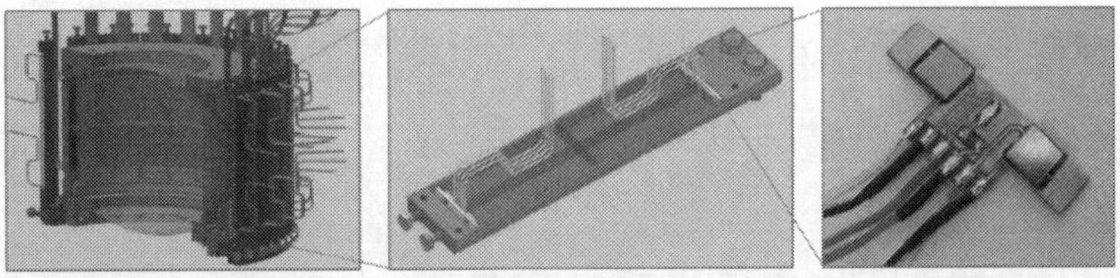

Figure 7. *Left:* Target cell surrounded by the SDDs, with the Veto-2 system mounted directly behind. *Middle:* Veto-2 detector unit consisting of four plastic scintillators read out by SiPMs. *Right:* The zoom shows the SiPMs in detail, with the LED for calibration in the middle.

3 Status of the Experiment

To establish the optimal working conditions for the SIDDHARTA-2 experiment, a run with a reduced setup, called SIDDHARTINO, was performed in 2021 during the commissioning operations of DAΦNE. With only eight out of the 48 SDD arrays installed, the optimal conditions for the SIDDHARTA-2 experiment were established. During this phase, the performance of all components of the setup, i.e. the luminometer, kaon trigger, cooling system, degrader, target cell and SDDs, was thoroughly checked. Moreover, the background conditions were studied - a crucial step for the planned deuterium measurement. During the entire SIDDHARTINO run, the online feedback provided by the luminosity monitor was vital in assessing the quality of the beam and the level of background.
In the summer of 2021, the commissioning phase was concluded with a ^{4}He run, for an integrated luminosity of 30.9 pb^{-1}. Data were collected for different ^{4}He gas densities (0.75% and 1.5% liquid He density) and various degrader configurations. A preliminary calibrated spectrum obtained with the ^{4}He target for 9.3 pb^{-1} is shown in Fig. 8. A 2.5 μs coincidence with the kaon signals registered by the kaon trigger was required. The kaonic ^{4}He L_α peak is clearly visible at 6.4 keV; additional peaks are present due to the interaction of the kaons with the setup materials.

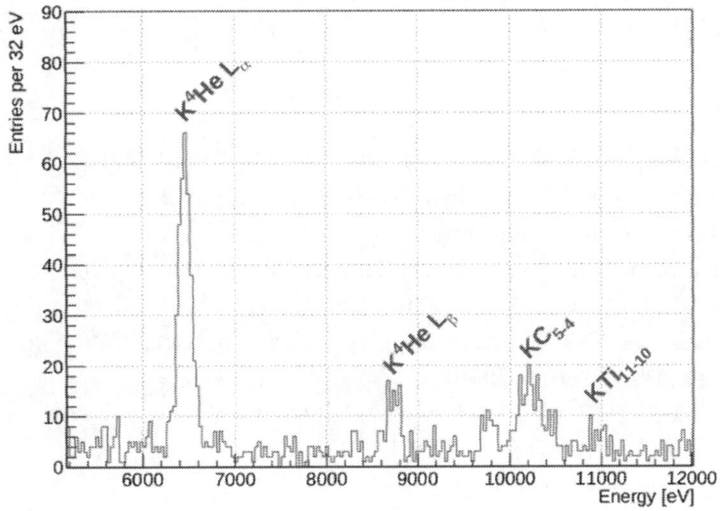

Figure 8. Preliminary calibrated spectrum obtained during the SIDDHARTINO run with ^{4}He for an integrated luminosity of 9.3 pb^{-1}.

After the successful conclusion of the SIDDHARTINO run, the full SIDDHARTA-2 setup will be installed, including full veto systems as well as all 48 SDD arrays. The collaboration aims to collect a total integrated luminosity of 800 pb^{-1} of kaonic deuterium data in 2022, split into two runs of 300 pb^{-1} and 500 pb^{-1}, respectively.

4 Conclusion and Future Perspectives

The experimental apparatus of SIDDHARTA-2 has been updated from the SIDDHARTA experiment to sufficiently increase the signal-to-background ratio for a successful kaonic deuterium measurement. The main updates include a lightweight gas target cell, new monolithic Silicon Drift Detectors and an active two-stage veto system for background suppression. All of these components were characterised and their performances were proven suitable for the planned kaonic deuterium measurement. A first successful run with kaonic ^{4}He was performed with a reduced setup during the DAΦNE commissioning phase. The full setup is currently under installation, and the kaonic deuterium run will be performed in 2022. In parallel, new proposals following the K^-d measurement by SIDDHARTA-2 are already in preparation. These include not only the development of new detectors, for example Cd(Zn)Te or High Purity Germanium detectors, but also the exploration of other elements like Be, B, or Pb, as well as taking advantage of the ultra-high precision of Highly Annealed Pyrolitic Graphite (HAPG) spectrometers.

This work was partly supported by the Austrian Science Fund (FWF): Doctoral program No. W1252-N27, as well as P24756-N20 and P33037-N; the Croatian Science Foundation under the project IP-2018-01-8570; EU STRONG-2020 project (grant agreement No. 824093), the Polish Ministry of Science and Higher Education grant No. 7150/E-338/M/2018 and the Foundational Questions Institute and Fetzer Franklin Fund, a donor advised fund of Silicon Valley Community Foundation (Grant No. FQXi-RFP-CPW-2008).
The authors acknowledge C. Capoccia from INFN-LNF and H. Schneider, G. Kaiser, L. Stohwasser, and D. Pristauz Telsnigg from Stefan-Meyer-Institut für Subatomare Physik for their fundamental contribution in designing and building the SIDDHARTA-2 setup. We would also like to thank the DAΦNE staff for the excellent working conditions and permanent support.

References

[1] C. Curceanu *et al.*, Rev. Mod. Phys. **91**, 025006 (2019).
[2] Y. Ikeda, T. Hyodo and W. Weise, Nucl. Phys. A **881**, 98-114 (2012).
[3] A. Cieplý and J. Smejkal, Nucl. Phys. A **881**, 115-126 (2012).
[4] M. Mai and U. G. Meißner, Eur. Phys. J. A **51**, 30 (2015).
[5] Z.-H. Guo and J. A. Oller, Phys. Rev. C **87**, 035202 (2013).
[6] A. Feijoo, V. Magas and A. Ramos, Phys. Rev. C **99**, 035211 (2019).
[7] A. Cieplý *et al.*, AIP Conference Proceedings **2249**, 030014 (2020).
[8] M. Bazzi *et al.*, Phys. Lett. B **704(3)**, 113-117 (2011).
[9] M. Bazzi *et al.*, Nucl. Phys. A **907**, 69 (2013).
[10] M. Skurzok *et al.*, J. Inst. **15**, P10010(1-13) (2020).
[11] M. Miliucci *et al.*, Condens. Matter **4**, 31(1-8) (2019).
[12] M. Iliescu *et al.*, IEEE Trans. Instrum. Meas. **70**, 9507807 (2021).
[13] M. Miliucci *et al.*, Meas. Sci. Techn. **32**, 095501(1-7) (2021).
[14] M. Bazzi *et al.*, J. Inst. **8**, T11003 (2013).
[15] M. Tüchler *et al.*, J. Phys. Conf. Ser. **1138**, 012012(1-8) (2018).

EPJ Web of Conferences **262**, 01017 (2022) https://doi.org/10.1051/epjconf/202226201017
EXA2021

Antinucleon-nucleon interaction from chiral effective field theory and its application to neutron-antineutron oscillations

Johann Haidenbauer[1,*] and *Ulf-G.* Meißner[2,1,**]

[1]Institute for Advanced Simulation, Institut für Kernphysik and Jülich Center for Hadron Physics, Forschungszentrum Jülich, D-52425 Jülich, Germany
[2]Helmholtz-Institut für Strahlen- und Kernphysik and Bethe Center for Theoretical Physics, Universität Bonn, D-53115 Bonn, Germany

Abstract. An $\bar{N}N$ potential is reviewed which has been derived within chiral effective field theory by the Jülich-Bonn group. By construction it reproduces $\bar{N}N$ phase shifts and inelasticities, provided by an elaborate phase-shift analysis of available $\bar{p}p$ scattering data. This interaction is then employed in a study of neutron-antineutron oscillations in the deuteron. In particular, the deuteron lifetime is considered, and results for it are presented in terms of the free-space $n - \bar{n}$ oscillation time, based on that $\bar{N}N$ potential in combination with an NN interaction likewise derived within chiral effective field theory.

1 Introduction

In Ref. [1] a new generation of nucleon-nucleon (NN) potentials was derived in the framework of chiral effective field theory (EFT). In that publication a novel local regularization scheme was introduced and applied to the pion-exchange contributions to the NN force. Furthermore, an alternative scheme for estimating the theoretical uncertainty was proposed that no longer depends on a variation of the cutoff [2, 3]. In a recent study the Jülich-Bonn group took advantage of this development and adapted the formalism and the concepts proposed by Epelbaum et al. in order to establish a high-precision antinucleon-nucleon ($\bar{N}N$) potential within chiral EFT [4]. Specifically, in that work an $\bar{N}N$ potential up to next-to-next-to-next-to-leading order (N^{3}LO) in the perturbative expansion was presented, which extends a previous work by the Jülich-Bonn group that had considered the $\bar{N}N$ force only up to N^{2}LO [5]. In both cases the strength parameters of the contact terms that arise in the EFT framework [3, 6] (i.e. the low-energy constants or simply LECs) have been fixed by a fit to the phase shifts and inelasticities provided by a proper phase-shift analysis of $\bar{p}p$ scattering data [7].

The $\bar{N}N$ potentials from chiral EFT have been already successfully used in studies of the electromagnetic form factors of the proton in the time-like region [8] and of the unusually large near-threshold enhancement in the $\bar{p}p$ spectrum detected in the reaction $J/\psi \to \gamma\bar{p}p$ [9]. A further and certainly more exotic application of this $\bar{N}N$ interaction has been presented by us in Ref. [10], where neutron-antineutron oscillations in the deuteron have been studied.

*e-mail: j.haidenbauer@fz-juelich.de
**e-mail: meissner@hiskp.uni-bonn.de

Neutron-antineutron ($n - \bar{n}$) oscillations involve a change of the baryon number (B) by two units ($|\Delta B| = 2$). An experimental observation would allow a glimpse on physics beyond the standard model, see e.g. [11–13]. Since in such oscillations B is violated the process satisfies one of the Sakharov conditions [14] that have been formulated in order to explain the observation that there is more matter than anti-matter in the universe [15].

The key quanitity in this subject is the free $n-\bar{n}$ oscillation time, $\tau_{n-\bar{n}}$. The presently best experimental limit on it is $\tau_{n-\bar{n}} > 0.86 \times 10^8$ s $\approx$ 2.7 yr (with 90 % C.L.) [16]. Additional information can be deduced from studies of $n - \bar{n}$ oscillations in a nuclear environment. Corresponding experiments have been performed, e.g., for ^{56}Fe [17], ^{16}O [18], and for the deuteron (^{2}H) [19], while others are planned [20, 21] In such a case the oscillation process is suppressed as compared to the free situation. The pertinent lifetime τ_{nuc} is commonly expressed in terms of the one in free space as [12]

$$\tau_{\text{nuc}} = R\,\tau^2_{n-\bar{n}}\,, \tag{1}$$

where R is an intranuclear suppression factor, also called reduced lifetime, that depends on the specific nucleus. It can be calculated from nuclear theory and then can be used to relate the measured lifetimes of those nuclei with the free $n - \bar{n}$ oscillation time [12], see, e.g., Refs. [22–26].

For a long time the suppression factors published in 1983 [23] have been used as standard by experimentalists in the interpretation of their measurements [17, 19]. For example, in case of the deuteron the corresponding value is $R \sim (2.40 - 2.56) \times 10^{22}$ s^{-1}, a prediction based on the phenomenological antinucleon-nucleon ($\bar{N}N$) potentials by Dover and Richard [27, 28]. Recently, however, those values have been challenged in a work by Oosterhof et al. [29]. In that study an effective field theory for the $|\Delta B| = 2$ interaction is constructed and the quantity R is evaluated within the power counting scheme proposed by Kaplan, Savage, and Wise (KSW) [30, 31] for the nucleon-nucleon (NN) and $\bar{N}N$ interactions. The value of R for the deuteron obtained in that approach is $(1.1 \pm 0.3) \times 10^{22}$ s^{-1}, about a factor 2 smaller than the one by Dover et al. [23].

In the light of this controversal situation a new calculation of the suppression factor for the deuteron has been performed by us [10]. It was prompted by the aim to utilize the discussed modern chiral interactions for the involved NN [1] and $\bar{N}N$ [4] systems. Specifically, in case of the $\bar{N}N$ interaction most of the available precise $\bar{p}p$ scattering data (mostly from the LEAR facility at CERN [32]) have appeared only after the publication of the potentials used in Ref. [23]. Therefore, an update is long overdue. Of course, the main motivation was the aforementiond discrepancy reported in Ref. [29] and the prospect to find a plausible explanation for that difference.

The paper is structured in the following way: In Sect. 2 the employed $\bar{N}N$ potential is described and some results for $\bar{n}p$, relevant for the calculation of $n - \bar{n}$ oscillations in the deuteron, are presented. In Sect. 3 a basic description of the formalism for evaluating the $n - \bar{n}$ oscillations in the deuteron is provided. Results for the oscillations, specifically for the suppression factor R, are presented in Sect. 4. The paper closes with a brief summary.

2 The $\bar{N}N$ interaction in chiral EFT

The derivation of the chiral $\bar{N}N$ potential using the Weinberg power counting is described extensively in Ref. [4], and we refer the reader to this work for the details. As already indicated above, the chiral potential contains pion exchanges and a series of contact interactions with an increasing number of derivatives. Up to N^{3}LO there are contributions from one-, two- and three-pion exchanges. Those are identical to the ones that appear in the NN potential

[1]. However, there is a sign change in case of an odd number of exchanged pions due to its negative G parity, i.e. $V^{\bar{N}N} = -V_\pi + V_{2\pi} - V_{3\pi}$, with V_π, etc., the corresponding contributions to the NN force. On the other hand, the contact interaction, $V^{\bar{N}N}_{cont}$, cannot be taken over simply from the NN case. Those contact terms represent effectively the short-range part of the interaction and, therefore, the G parity of the indivudal contributions remains unresolved. Thus, the strength parameters associated with the arising contact terms, the LECs, need to be determined in a fit to $\bar{N}N$ data. We fix them by fitting to the phase shifts and inelasticity parameters of the PWA of Zhou and Timmermans [7]. How this is done is described in detail in Ref. [4]. Note that there are more independent LECs in the $\bar{N}N$ case than in NN because in the former system there is no restriction from the Pauli principle. In addition, in our approach $\bar{N}N$ annihilation (into multi-meson channels) is likewise parameterized by contact terms, see Refs. [4, 5] for explicit expressions.

Once the potential is established, the reaction amplitude is obtained from the solution of a relativistic Lippmann-Schwinger (LS) equation. It reads in partial-wave projected form

$$T_{L''L'}(p'',p';E_k) = V_{L''L'}(p'',p') + \sum_L \int_0^\infty \frac{dpp^2}{(2\pi)^3} V_{L''L}(p'',p)\frac{1}{2E_k - 2E_p + i0^+} T_{LL'}(p,p';E_k)\,. \quad (2)$$

Here p'' and p' are the (moduli of the) center-of-mass $\bar{N}N$ momenta in the final and initial states, respectively, and $E_k = \sqrt{M_p^2 + k^2}$, where k is the on-shell momentum. We adopt a relativistic scattering equation so that our amplitudes fulfill the relativistic unitarity condition at any order, as done also in the NN sector [2, 6]. On the other hand, relativistic corrections to the potential are calculated order by order. They appear first at N^{3}LO in the Weinberg scheme, see Appendix A in Ref. [4]. L, L', etc., specifies the orbital angular momentum, considering that the $\bar{N}N$ system can be in uncoupled (spin-singlet or triplet) states where $L'' = L' = L = J$ or in coupled partial waves with $L'', L', L = J-1, J+1$.

Since the integral in the LS equation (2) is divergent for the chiral potentials [2, 6] a regularization needs to be introduced. For that the regularization scheme of Ref. [1] is utilized, where a local regulator is used for the pion-exchange contributions and a nonlocal regulator for the contact terms:

$$\begin{aligned} V_{n\pi}(q) &\to V_{n\pi}(r) \times f_R(r) \to V^{reg}_{n\pi}(q); \qquad (\vec{q} = \vec{p}\,' - \vec{p}) \\ V_{cont} &= V(p',p) \to V(p',p) \times f_\Lambda(p',p) = V^{reg}_{cont}\,. \end{aligned} \quad (3)$$

The explict form of the regulation functions is $f_R(r) = \left[1 - exp(-r^2/R^2)\right]^6$ and $f_\Lambda(p,p') = exp(-(p'^2 + p^2)/\Lambda^2)$. The cutoff radius R is varied in the range R = 0.7-1.2 fm [1] where $\Lambda = 2/R$ is used for relating the momentum-space cutoff parameter with the cutoff radius.

A complete overview of our results for $\bar{N}N$ scattering up to N^{3}LO in chiral EFT can be found in Ref. [4] while possible $\bar{N}N$ bound states are discussed in Ref. [33]. Here we focus on quantities that are relevant for the study of $n - \bar{n}$ oscillations in the deuteron. Specifically, we look at the $^3S_1$-$^3D_1$ phase shifts in the $\bar{n}p$ (isospin I = 1) channel (Fig. 1) and the $\bar{n}p$ cross section, see Fig. 2. Note that data for the latter reaction were not included in the PWA [7] because they are less precise than the ones for $\bar{p}p$. The bands in Figs. 1 and 2 represent the estimated uncertainty of our results. Here we follow the suggestion of Ref. [1] and use as measure the expected size of the higher-order corrections together with the actual size of the higher-order corrections, see Ref. [4] for details.

In Fig. 3 we present the $\bar{n}p$ annihilation cross section multiplied by the velocity β of the incoming $\bar{n}$. This allows a more detailed view on the situation very close to the threshold. We consider only the N^{3}LO interaction with cutoff radius R = 0.9 fm because the theoretical

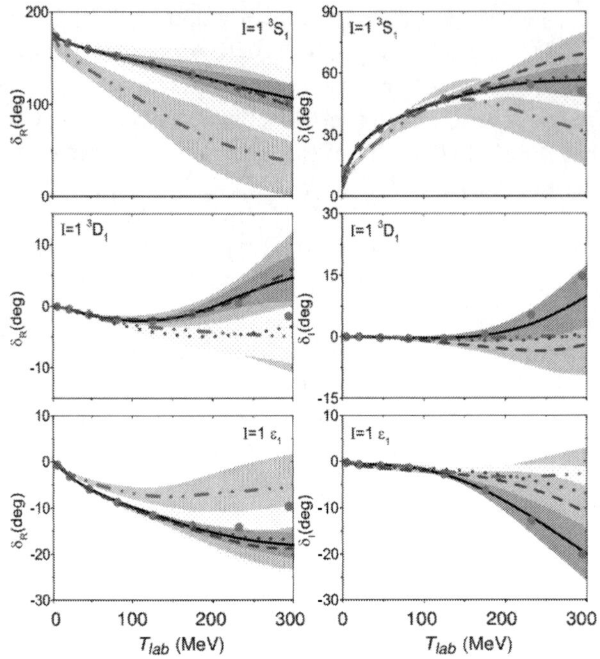

Figure 1. Real and imaginary parts of the phase shifts in the $^3S_1$-$^3D_1$ partial wave with isospin $I = 1$. Results at N^{3}LO (black solid line), N^{2}LO (blue dashed line), and NLO (red dotted line) are shown. The estimated uncertainty is indicated by bands: N^{3}LO (magenta), N^{2}LO (cyan), and NLO (yellow). The filled circles represent the solution of the $\bar{p}p$ PWA [7].

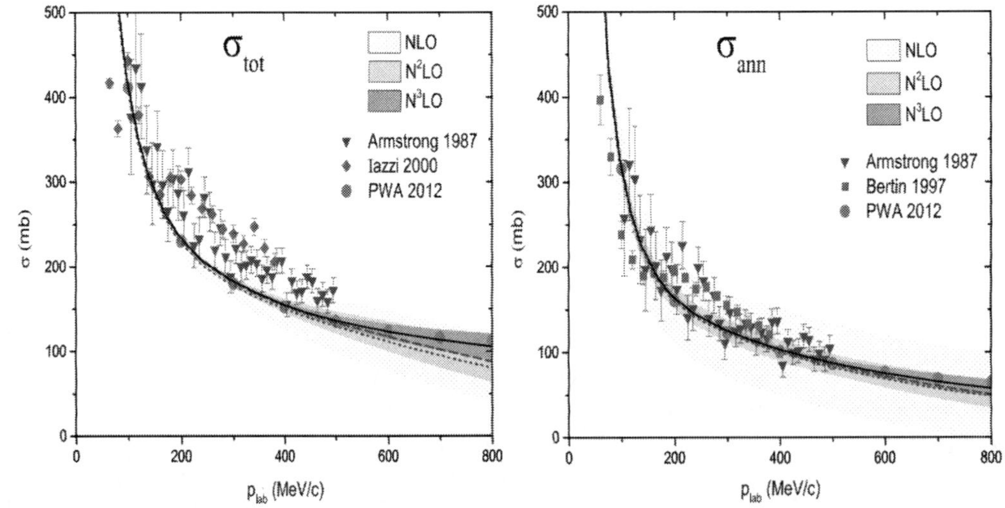

Figure 2. Total (σ_{tot}) and integrated annihilation (σ_{ann}) cross sections for $\bar{n}p$ scattering. Same description of curves and bands as in Fig. 1. Data are from Refs. [34–36].

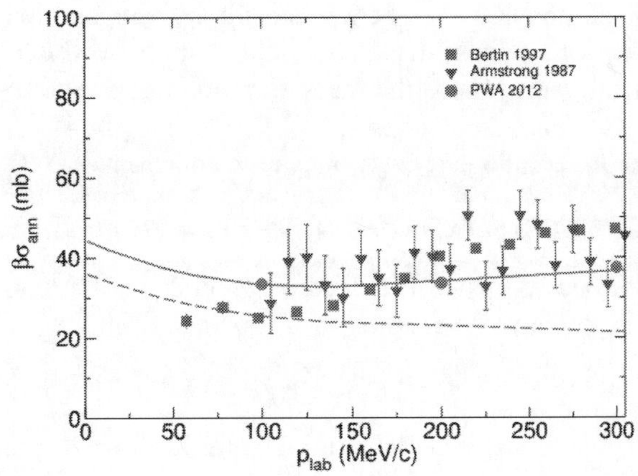

Figure 3. Annihilation cross section for $\bar{n}p$ scattering multiplied with the velocity β of the incoming antineutron. The curves show the results for the N^{3}LO interaction with cutoff radius $R = 0.9$ fm [4]. The dashed line is the contribution from the $^3S_1$-$^3D_1$ partial wave while the solid line is the full result. Data are from Refs. [34, 36].

uncertainty at low momenta is very small. One can see that most of the contributions come indeed from the $^3S_1$-$^3D_1$ partial wave, that is relevant for the evaluation of $n-\bar{n}$ oscillations. The figure reveals also that the data from the OBELIX Collaboration [36] do not follow the expected near-threshold behavior. The cross sections for exothermic reactions should behave like $1/\beta$ so that $\beta\sigma_{ann}$ is then approximately constant for low momenta. This expectation is well satisfied by the BNL data [34], but also by the theory predictions. A discussion of the anomalous behavior suggested by the OBELIX data can be found in Ref. [35].

In the study of $n-\bar{n}$ oscillations below we employ also an earlier (N^{2}LO) $\bar{N}N$ potential published by our group. A description of that interaction can be found in Ref. [5]. We consider the two $\bar{N}N$ interactions because they are based on rather different regularization schemes. In the earlier potential [5] a non-local exponential exponential regulator was employed for the whole potential while, as outlined above, in the N^{3}LO interaction [4] a local regulator was adopted for the evaluation of the one- and two-pion contributions. Comparing the pertinent results allows us to shed light on the question in how far the choice of the regulator influences the predictions.

For the needed deuteron wave function we take those from the NN potentials derived in the same framework and at the same order. In particular, for the calculation with the new $\bar{N}N$ interaction [4] we take the wave functions from Ref. [1], whereas for that with the (N^{2}LO) $\bar{N}N$ potential [5] the wave functions from Ref. [2] are utilized. For exploring the sensitivity of the results to the deuteron wave function we employ also those from two meson-exchange potentials [37, 38].

3 Formalism for $n-\bar{n}$ oscillations

For evaluating the $n-\bar{n}$ oscillations in the deuteron we follow the formalism presented in Refs. [22, 23]. The starting point is the eigenvalue (Schrödinger) equation [22]

$$\begin{pmatrix} H_0 + V_{np} & V_{n-\bar{n}} \\ V_{n-\bar{n}} & H_0 + V_{\bar{n}p} \end{pmatrix} \begin{pmatrix} |\psi_{np}\rangle \\ |\psi_{\bar{n}p}\rangle \end{pmatrix} = (E - i\Gamma/2) \begin{pmatrix} |\psi_{np}\rangle \\ |\psi_{\bar{n}p}\rangle \end{pmatrix} . \quad (4)$$

Here, V_{np} and $V_{\bar{n}p}$ are the potentials in the np and $\bar{n}p$ systems and $|\psi_{np}\rangle$ and $|\psi_{\bar{n}p}\rangle$ are the corresponding wave functions. The systems are coupled via $V_{n-\bar{n}}$ which is given by the $n-\bar{n}$ transition matrix element $\delta m_{n-\bar{n}}$ where the latter is proportional to the inverse of the $n-\bar{n}$ oscillation time, i.e. $V_{n-\bar{n}} = \delta m_{n-\bar{n}} = \hbar/\tau_{n-\bar{n}}$ [12].

To leading order the $\bar{n}p$ component $|\psi_{\bar{n}p}\rangle$ obeys the equation

$$(H_0 + V_{\bar{n}p} - E_d)|\psi_{\bar{n}p}\rangle = -V_{n-\bar{n}}|\psi_d\rangle\,, \tag{5}$$

where E_d is the unperturbed energy of the deuteron and $|\psi_d\rangle$ is the corresponding deuteron wave function. The decay width of the deuteron, Γ_d, is then [22]

$$\Gamma_d = -2\,V_{n-\bar{n}}\,\mathrm{Im}\langle\psi_d|\psi_{\bar{n}p}\rangle\,. \tag{6}$$

We solve Eq. (5) in momentum space. Performing a partial wave decomposition and taking into account the coupling of the $^3S_1$ and $^3D_1$ channels, the above integral equation reads

$$(2E_p - E_d)\,\psi^L_{\bar{n}p}(p) + \sum_{L'}\int \frac{dq\,q^2}{(2\pi)^3} V^{L,L'}_{\bar{n}p}(p,q)\,\psi^{L'}_{\bar{n}p}(q) = -V_{n-\bar{n}}\,\psi^L_d(p)\,, \tag{7}$$

with $L,\,L' = 0,\,2$. Note that E_d is the total energy corresponding to the deuteron, i.e. $E_d - 2m_N = 2\sqrt{m_N^2 - \kappa^2} - 2m_N = -B_d$ where B_d is the standard binding energy of 2.225 MeV and $\kappa = \sqrt{m_N B_d} \simeq 45.7$ MeV is the binding momentum. The deuteron wave function is normalized by

$$\int dpp^2\left[(\psi^0_d(p))^2 + (\psi^2_d(p))^2\right] = 1\,, \tag{8}$$

and the width is provided by

$$\Gamma_d = -2\,V_{n-\bar{n}}\,\mathrm{Im}\sum_L \int dpp^2\,\psi^L_d(p)\,\psi^L_{\bar{n}p}(p)\,. \tag{9}$$

The deuteron lifetime τ_d is given by $\tau_d = \hbar/\Gamma_d$. The interesting quantity is the so-called reduced lifetime R [22, 23, 26] which relates the free $n-\bar{n}$ oscillation lifetime with that of the deuteron,

$$\tau_d = R\,\tau^2_{n-\bar{n}} \quad \text{i.e.} \quad R = \frac{\hbar}{\Gamma_d \tau^2_{n-\bar{n}}}\,. \tag{10}$$

4 Results and discussion

Our results for the reduced lifetime R for the deuteron are summarized in Table 1. They are based on our N^{3}LO interaction with cutoff $R_0 = 0.9$ fm [4] and the N^{2}LO interaction with cutoff $\{\Lambda, \tilde{\Lambda}\} = \{450, 500\}$ MeV [5]. For technical details see the corresponding publications. Besides the predictions for R based on the chiral $\bar{N}N$ interactions we list also the values given in Ref. [23] where the $\bar{N}N$ potentials DR$_1$ and DR$_2$ by Dover-Richard [27, 28] have been utilized. Furthermore we include results from the calculation of Oosterhof et al. performed directly within EFT on the basis of the KSW approach. In this case R can be represented in a compact analytical form which reads up to NLO [29]

$$R = -\frac{\kappa}{m_N}\,\frac{1}{\mathrm{Im}\,a_{\bar{n}p}}\,\frac{1}{1 + 0.4 + 2\kappa\,\mathrm{Re}\,a_{\bar{n}p} - 0.13 \pm 0.4}\,. \tag{11}$$

Obviously, the only parameter here is the $\bar{n}p$ $^3S_1$ scattering length. All other quantities that enter are well established NN observables, cf. Ref. [29] for details. Note that in that paper, the scattering length $\mathrm{Re}\,a_{\bar{n}p}$ was taken from Ref. [4].

Table 1. Reduced lifetime R calculated for the χEFT $\bar{N}N$ potentials from Refs. [4, 5], together with information on the pertinent $\bar{n}p$ $^3S_1$ scattering length. Results for the Dover-Richard potentials DR_1 and DR_2 are taken from Ref. [23]. The corresponding scattering lengths are from Ref. [39]. Predictions based on Eq. (11), i.e. on the KSW approach applied in Ref. [29], are indicated too.

	χEFT N^{2}LO [5]	χEFT N^{3}LO [4]	DR_1 [23]	DR_2 [23]
R [s^{-1}]	2.49×10^{22}	2.56×10^{22}	2.56×10^{22}	2.40×10^{22}
(Eq. (11))	$(1.1 \pm 0.3) \times 10^{22}$	$(1.2 \pm 0.3) \times 10^{22}$	$(1.4 \pm 0.4) \times 10^{22}$	$(1.3 \pm 0.3) \times 10^{22}$
$a_{^3S_1}$ [fm]	$0.44 - \mathrm{i}\,0.91$	$0.44 - \mathrm{i}\,0.96$	$0.87 - \mathrm{i}\,0.66$	$0.89 - \mathrm{i}\,0.71$

Table 1 reveals that the values for R predicted by the chiral $\bar{N}N$ interactions are fairly similar to those obtained for the DR potentials in the past. By contrast, the results based on the framework employed by Oosterhof et al. [29] are rather different. Since the scattering length from the N^{3}LO chiral $\bar{N}N$ interaction [4] is utilized in that work, the large discrepancy observed in Ref. [29] is certainly not due to differences in Im $a_{\bar{n}p}$ but must be primarily a consequence of the different approaches.

For investigating the sensitivity of our results to the used ingredients we performed various exploratory calculations. Specifically, we employed the NLO and N^{2}LO variants of the considered $\bar{N}N$ (and NN) interactions. The corresponding predictions for R were found to lie within a range of $(2.48 - 2.65) \times 10^{22}$ s^{-1}. If one takes this variation as measure for the uncertainty due to the nuclear structure, i.e. the NN and $\bar{N}N$ interactions (wave functions), a value of roughly $R = (2.6 \pm 0.1) \times 10^{22}$ s^{-1} can be deduced. Application of the method proposed in Ref. [1] for estimating the uncertainty to the calculation based on the $\bar{N}N$ interaction from 2017 [4], say, leads to a slightly smaller uncertainty. We have also varied the deuteron wave functions alone. As an extreme case we even took wave functions from phenomenological NN potentials derived in an entirely different framework, namely in the meson-exchange picture [37, 38]. Also here the obtained values for R remained within the range given above. Finally, omitting the D-wave component of the deuteron wave function in our calculation causes a 5 % variation. It leads to an increase of R and, thus, does not bring the result closer to the values presented by Oosterhof et al. Overall, we confirm the observation by Dover et al. that the predictions for R are fairly insensitive to the details of the employed $\bar{N}N$ potentials [23], provided that these potentials describe the $\bar{p}p$ data at low energies.

At this stage we do not have a sensible explanation for the difference of our results (and those of Ref. [23]) to the ones of Oosterhof et al. [29]. However, we believe that it is due to the fact that in the latter work the width Γ_d is evaluated following the perturbative scheme developed by Kaplan, Savage, and Wise [31]. In that scheme there is no proper deuteron wave function. Rather one works with an effectively constructed wave function that is represented in terms of an irreducible two-point function [29, 31]. This seems to work well for some electromagnetic form factors of the deuteron, at least at low momentum transfer [31, 40]. On the other hand, the quadrupole moment of the deuteron is overestimated by 40 % [31], which hints that the properties of the wave function at large distances (small momenta) are not that well represented in this scheme. Clearly, this should have an impact on the quantity studied in the present work as well. Note that a comparable agreement (mismatch) with regard to the KSW scheme has been also observed in studies of the electric dipole moment (magnetic quadrupole moment) of the deuteron [41–43]. In any case, one should not forget that there is convergence problem of the KSW approach for NN partial waves where the tensor force from pion exchange is present [44]. It affects specifically the $^3S_1$-$^3D_1$ channel where difficulties appear already for momenta around 100 MeV/c, see [44] and also the discussions in Refs. [3, 45].

5 Summary

Chiral effective field theory does not only work very well for the interaction between nucleons [3, 6] it can be also successfully employed to study the antinucleon-nucleon interaction as has been demonstrated by the Jülich-Bonn group [4, 5]. In this contribution we have discussed our most refined $\bar{N}N$ potential [4], established within chiral EFT up to N^{3}LO, and we presented some selective predictions for $\bar{n}p$ scattering. The potential has been fitted to up-to-date $\bar{N}N$ phase shifts and inelasticities provided by a phase-shift analysis of available $\bar{p}p$ scattering data [7].

As an application of this interaction neutron-antineutron oscillations in the deuteron have been considered. In particular, results for the deuteron lifetime have been presented, evaluated in terms of the free-space $n-\bar{n}$ oscillation time, utilizing that $\bar{N}N$ potential in combination with an NN interaction likewise derived within chiral effective field theory. The value obtained for the so-called reduced lifetime R which relates the free-space $n-\bar{n}$ oscillation time $\tau_{n-\bar{n}}$ with the deuteron lifetime is found to be $R = (2.6 \pm 0.1) \times 10^{22}\ \mathrm{s}^{-1}$, where the quoted uncertainty is due to the NN and $\bar{N}N$ interactions (wave functions).

Our prediction for R agrees with the value obtained by Dover and collaborators almost four decades ago [23] but deviates from recent EFT calculations, based on the perturbative scheme proposed by Kaplan, Savage, and Wise [29], by about a factor of 2. A possible explanation for the difference could be that the KSW scheme does not involve a proper deuteron wave function. Rather this ingredient is represented effectively in terms of an irreducible two-point function. It is known from past studies that the KSW approach fails to describe quantities that depend more sensitively on the wave function like, for example, the quadrupole moment of the deuteron [31].

Acknowledgements

This work is supported in part by the DFG and the NSFC through funds provided to the Sino-German CRC 110 "Symmetries and the Emergence of Structure in QCD" (DFG grant. no. TRR 110) and the VolkswagenStiftung (grant no. 93562). The work of UGM was supported in part by The Chinese Academy of Sciences (CAS) President's International Fellowship Initiative (PIFI) (grant no. 2018DM0034).

References

[1] E. Epelbaum, H. Krebs and U.-G. Meißner, Eur. Phys. J. A **51**, 53 (2015).
[2] E. Epelbaum, W. Glöckle and U.-G. Meißner, Nucl. Phys. A **747**, 362 (2005).
[3] E. Epelbaum, H.-W. Hammer and U.-G. Meißner, Rev. Mod. Phys. **81**, 1773 (2009).
[4] L. Y. Dai, J. Haidenbauer and U.-G. Meißner, JHEP **1707**, 078 (2017).
[5] X. W. Kang, J. Haidenbauer and U.-G. Meißner, JHEP **1402**, 113 (2014).
[6] R. Machleidt and D. R. Entem, Phys. Rept. **503**, 1 (2011).
[7] D. Zhou and R. G. E. Timmermans, Phys. Rev. C **86**, 044003 (2012).
[8] J. Haidenbauer, X.-W. Kang and U.-G. Meißner, Nucl. Phys. A **929**, 102 (2014).
[9] X.-W. Kang, J. Haidenbauer and U.-G. Meißner, Phys. Rev. D **91**, 074003 (2015).

[10] J. Haidenbauer and U.-G. Meißner, Chin. Phys. C **44**, 033101 (2020).
[11] R. N. Mohapatra, J. Phys. G **36**, 104006 (2009).
[12] D. G. Phillips II *et al.*, Phys. Rept. **612**, 1 (2016).
[13] S. Gardner and X. Yan, Phys. Lett. B **790**, 421 (2019).
[14] A.D. Sakharov, JETP Lett. **5**, 24 (1967).
[15] C. Grojean, B. Shakya, J. D. Wells and Z. Zhang, Phys. Rev. Lett. **121**, 171801 (2018).
[16] M. Baldo-Ceolin *et al.*, Z. Phys. C **63**, 409 (1994).
[17] J. Chung *et al.*, Phys. Rev. D **66**, 032004 (2002).
[18] K. Abe *et al.* [Super-Kamiokande], Phys. Rev. D **103**, 012008 (2021).
[19] B. Aharmim *et al.* [SNO], Phys. Rev. D **96**, 092005 (2017).
[20] J. L. Barrow, E. S. Golubeva, E. Paryev and J.-M. Richard, Phys. Rev. D **101**, 036008 (2020).
[21] A. Addazi *et al.*, J. Phys. G **48**, 070501 (2021).
[22] P. G. H. Sandars, J. Phys. G **6**, L161 (1980).
[23] C. B. Dover, A. Gal and J.-M. Richard, Phys. Rev. D **27**, 1090 (1983).
[24] W. M. Alberico, A. De Pace and M. Pignone, Nucl. Phys. A **523**, 488 (1991).
[25] J. Hufner and B. Z. Kopeliovich, Mod. Phys. Lett. A **13**, 2385 (1998).
[26] E. Friedman and A. Gal, Phys. Rev. D **78**, 016002 (2008).
[27] C. B. Dover and J.-M. Richard, Phys. Rev. C **21**, 1466 (1980).
[28] C. B. Dover and J.-M. Richard, Phys. Rev. C **25**, 1952 (1982).
[29] F. Oosterhof, B. Long, J. de Vries, R. G. E. Timmermans and U. van Kolck, Phys. Rev. Lett. **122**, 172501 (2019).
[30] D. B. Kaplan, M. J. Savage and M. B. Wise, Phys. Lett. B **424**, 390 (1998).
[31] D. B. Kaplan, M. J. Savage and M. B. Wise, Phys. Rev. C **59**, 617 (1999).
[32] E. Klempt, F. Bradamante, A. Martin and J.-M. Richard, Phys. Rept. **368**, 119 (2002).
[33] J. Haidenbauer, EPJ Web Conf. **181**, 01028 (2018).
[34] T. Armstrong et al., Phys. Rev. D **36**, 659 (1987).
[35] F. Iazzi et al., Phys. Lett. B **475**, 378 (2000).
[36] A. Bertin *et al.* [OBELIX Collaboration], Nucl. Phys. Proc. Suppl. **56**, 227 (1997).
[37] J. Haidenbauer, K. Holinde and M. B. Johnson, Phys. Rev. C **48**, 2190 (1993).
[38] R. Machleidt Phys. Rev. C **63**, 024001 (2001).
[39] J. Carbonell, J.-M. Richard and S. Wycech, Z. Phys. A **343**, 325 (1992).
[40] M. Walzl and U.-G. Meißner, Phys. Lett. B **513**, 37 (2001).
[41] J. de Vries, E. Mereghetti, R. G. E. Timmermans and U. van Kolck, Phys. Rev. Lett. **107**, 091804 (2011).
[42] C.-P. Liu, J. de Vries, E. Mereghetti, R. G. E. Timmermans and U. van Kolck, Phys. Lett. B **713**, 447 (2012).
[43] J. Bsaisou, J. de Vries, C. Hanhart, S. Liebig, U.-G. Meißner, D. Minossi, A. Nogga and A. Wirzba, JHEP **1503**, 104 (2015) Erratum: [JHEP **1505**, 083 (2015)].
[44] S. Fleming, T. Mehen and I. W. Stewart, Nucl. Phys. A **677**, 313 (2000).
[45] P. F. Bedaque and U. van Kolck, Ann. Rev. Nucl. Part. Sci. **52**, 339 (2002).

Antiproton-nuclei cross sections with Woods-Saxon potential at low energies

Andrea Bianconi[1,2], *Giovanni* Costantini[1,2], *Giulia* Gosta[1,2], *Marco* Leali[1,2], *Valerio* Mascagna[1,2], *Stefano* Migliorati[1,2,*], and *Luca* Venturelli[1,2]

[1]Dipartimento di Ingegneria dell'Informazione (DII), Università degli Studi di Brescia, Brescia, Italy
[2]INFN - Sezione di Pavia, Pavia, Italy

Abstract. The present knowledge of the antinucleons elastic scattering and annihilation processes in matter at low energies is limited to a few nuclei data in a small phase-space. Optical potential models are useful tools for modelling nuclear strong interaction of antinucleons with matter providing predictions at very low energies where data are missing. New calculations of elastic and annihilation cross sections for antiproton with nuclei using an optical potential of Woods-Saxon (WS) shape are presented. Preliminary predictions at low energies for carbon and calcium show clearly-measurable nuclear effects for nuclear elastic cross sections at large angles and momenta greater than 50 MeV/c. Some discrepancies in annihilation cross section comparing predictions and data are present using the same fitting parameters.

1 Introduction

During the 1980s and 1990s, experiments at Low Energy Antiproton Ring (LEAR) explored antinucleons annihilation and elastic scattering processes at low energy ($E < 50$ MeV) with different nuclei [1–16]. More recently, annihilation cross section with antiprotons of momentum $p = 100$ MeV/c (correponding to kinetic energy of $E = 5.3$ MeV) were measured by the ASACUSA collaboration [17–21] at the Antiproton Decelerator (AD) facility. In addition, other very low-energy measurements ($E = 125$ keV) were performed by the same collaboration on some nuclei, obtaining upper and lower limits of the values of annihilation cross sections [22]. The knowledge of the antinucleons-nucleus interaction at low energies is of interest for nuclear physics studies, in particular for the search of nuclear resonance and the determination of potential parameters. Futhermore, annihilation cross sections can also provide useful information about possible signals of antimatter presence in the Universe in the frame of some cosmological hypotheses [23, 24]. The knowledge of annihilation processes at low energies is also useful to study low-energy antiprotons from extra-atmospheric cosmic rays [25, 26] and also for medical physics applications [27–29]. Predictions at low energy can be used to design new experiments with enough sensitivity to study nuclear interaction effects and to observe other possible exotic processes.

In the present work, a six free-parameters Woods-Saxon (6pWS) optical potential has been chosen to model the strong interaction at low energies for the study of antiproton interaction with nuclei. With this optical potential, we are able to determine the parameters of the differential cross section for elastic scattering process of antiprotons with three nuclei of increasing mass: carbon-12 (^{12}C), calcium-40 (^{40}Ca) and lead-208 (^{208}Pb). We validate the model by comparing the annihilation cross sections, calculated with the obtained parameters, with the available data at low and intermediate momenta (p = 50–900 MeV/c). After this validation, we use the model to evaluate the elastic scattering differential cross sections at low momenta (with p = 15–100 MeV/c) considering in particular large angles (*i.e.* $\theta > 90°$).

The discussion is organized as follow. In Section 2, a brief review about optical potentials and partial wave expansion formalism is given; moreover, the choice of our model is presented and the fitting methods are briefly described. In Section 3, we present the main results of our analysis and a discussion about low-energy predictions of annihilation cross sections and on differential elastic scattering cross sections at large angles. In Section 4, a summary and possible future developments of the present analysis conclude the discussion.

2 Methods

To describe nuclear phenomena, a function for the nuclear strong potential generated by the target nucleus is defined. This potential depends on some parameters that describe its shape and on kinetics variables of both target and projectile. Strong potential of the system is usually modelled as a complex optical potential:

$$V_{opt}(r,k) = -\left[U(r,k) + iW(r,k)\right] \qquad (1)$$

where r is the radial position with respect to the center of the (spherical) nucleus and $k = p/\hbar$ is the wave number of the projectile with momentum p in the lab frame. The

*e-mail: s.migliorati016@unibs.it

real part of the potential accounts for nuclear elastic scattering processes, while the imaginary part represents inelastic processes. These inelastic processes, at low energy, are dominated by annihilation and charge exchange. Other processes (*e.g.* knockout reactions) contribute in minor percentage to the reaction cross section. From now on we will refer to reaction cross sections at low energy as annihilation cross sections, also above the threshold for which the charge exchange start to contribute, *i.e.* $E \sim 5\,\mathrm{MeV}$.
The shape of the optical potential can vary, depending on the used model. In such models, it is important to describe appropriately the surface of the nucleus. In fact, if the potential shape is step-like (*e.g.* square-well) the annihilation cross section results underestimated [30]. A Woods-Saxon (WS) function is frequently used to accomplish the requirement of "smoothed" surface. The 2-parameter WS (2pWS) function is defined by:

$$F_{WS}(r \mid r_0, a_0) = \left[1 + \exp\left(\frac{r - r_0}{a_0}\right)\right]^{-1} \qquad (2)$$

where r_0 is the nuclear radius, and a_0 is the diffuseness of the potential (the nuclear "surface thickness"). Therefore, the optical potential can be written as:

$$\begin{aligned} V_{opt}(r) = & -[U_0(k)\,F_{WS}(r \mid r_{0,U}, a_{0,U}) \\ & + iW_0(k)\,F_{WS}(r \mid r_{0,W}, a_{0,W})] \end{aligned} \qquad (3)$$

where U_0 and W_0 are the strengths of the real and imaginary parts of the potential.

The optical potential is then used in a differential equation which describes the system under analysis. In this case, the system is a projectile particle which collides with a nucleus target. To describe low-energy particles, a non-relativistic Schrödinger equation is a good approximation. Therefore, the projectile-target system is described by the following equation:

$$\left(\frac{\hbar^2}{2\mu}\nabla^2 - V_C(r) - V_{opt}(r)\right)\psi(\mathbf{r}) = 0 \qquad (4)$$

where V_C is the Coulomb potential of the projectile-target system:

$$V_C(r) = \begin{cases} \frac{Z_t Z_p e^2}{2R_C}\left(3 - \frac{r^2}{R_C^2}\right), & r \le R_C \\ \frac{Z_t Z_p e^2}{r}, & r > R_C. \end{cases} \qquad (5)$$

Z_t and Z_p are respectively the target and projectile charge in unit of elementary charge, e (in the antiproton case $Z_p = -1$) and R_C is the Coulomb radius that we assume to be $R_C = 1.25\,A^{1/3}\,\mathrm{fm}$ where A is the mass of the nucleus[1]. The equation is solved in partial-waves formalism, where the solution has the form:

$$\psi(r,\theta) \propto e^{ikz} + \frac{e^{ikr}}{r} f(\theta, k) \qquad (6)$$

with f the scattering amplitude of the outgoing wave and θ the angle between the direction of the incoming wave (along the z-axis) and the direction of the outgoing wave. This scattering amplitude is the sum of contributions from many partial-waves with different angular momentum quantum number l. The general expression for f is:

$$f(\theta, k) = \frac{1}{2ik}\sum_{l=0}^{l_{max}} (2l+1)(\bar{S}_l - 1)P_l(\cos\theta) \qquad (7)$$

where $\bar{S}_l$ is a function of l that contains the phase shift produced by the interaction under analysis. In principle, $l_{max} \to \infty$; however, for low energy projectiles and medium-light nuclei targets, only a limited number of partial-wave significantly contributes to the calculation. Considering both Coulomb and nuclear strong interactions acting in the processes, we define a total scattering amplitude:

$$f_{tot}(\theta, k) = f_C(\theta, k) + f_N(\theta, k) \qquad (8)$$

$$f_C(\theta, k) = \frac{1}{2ik}\sum_{l=0}^{l_{max}} (2l+1)(e^{2i\sigma_l} - 1)P_l(\cos\theta) \qquad (9)$$

$$f_N(\theta, k) = \frac{1}{2ik}\sum_{l=0}^{l_{max}} (2l+1)e^{2i\sigma_l}(S_l - 1)P_l(\cos\theta). \qquad (10)$$

f_C is the Coulomb-only scattering amplitude, assuming a point-like target-charge; f_N is the nuclear scattering amplitude, which include the influence of the Coulomb interaction of the nucleus; $\sigma_l = \arg[\Gamma(1 + l + i\eta)]$ is the Coulomb phase shift, where Γ is the Euler gamma function, $\eta = Z_t e^2 \mu/\hbar p$ is the Sommerfeld parameter, with μ the reduced mass of the projectile-target system. S_l is the nuclear factor which contains quantitative information about the interaction. The cross sections relate to f_{tot} and S_l by:

$$\frac{d\sigma_{el}}{d\Omega} = |f_{tot}(\theta)|^2 = \frac{1}{4k^2}\left|\sum_{l=0}^{l_{max}} (2l+1)(e^{2i\sigma_l}S_l - 1)P_l(\cos\theta)\right|^2 \qquad (11)$$

$$\sigma_{el}(k) = \frac{\pi}{k^2}\sum_{l=0}^{l_{max}} (2l+1)\left|e^{2i\sigma_l}S_l - 1\right|^2 \qquad (12)$$

$$\sigma_{an}(k) = \frac{\pi}{k^2}\sum_{l=0}^{l_{max}} (2l+1)(1 - |S_l|^2) \qquad (13)$$

where $d\sigma_{el}/d\Omega$ is the differential elastic cross section and σ_{an} is the annihilation cross section. The function f_N is calculated using a Fortran code derived from `THREEDEE` by Chant and Roos [31], which solves the Schrödinger equation in presence of the Coulomb potential and the 6pWS optical potential to model the strong interaction. The code calculations are up to $l = l_{max} = 120$. The differential elastic cross section and the other quantities are then computed for a specific momentum and angle.

Using this code, we fit to differential nuclear elastic cross section data minimizing a standard-χ^2. The parameters to be calculated are the six ones of the optical potential (real and imaginary parts) of Eqn 3. Standard algorithms of `ROOT` packages (*e.g.*, `TMinuit` library) [32] were used

[1] A is expressed in atomic mass unit (amu).

to find the minimum of the χ^2. We also estimate the 95% Confidence Interval (CI) band of the best fit lines, using:

$$CI_{95\%} = t_{NDF-1,\,0.95}\,\delta_{X,th} \tag{14}$$

where NDF are the number of degrees of freedom – number of data used for the fit minus the number of free-parameters –, t is the t-Student quantile and $\delta_{X,th}$ is the error estimation of the theoretical differential cross-section $X^{(th)}$ via error propagation.

3 Results and discussion

We fit our model to differential elastic cross section data of three different nuclei with increasing mass: ^{12}C, ^{40}Ca and ^{208}Pb. These data are from Garreta *et al.* (PS184 experiment at LEAR) [33] and Janouin *et al.* [34]. The results of the fitting procedure can be found in Table 1.

We have assumed $U_0(k) = U_0$ and $W_0(k) = W_0$. Also, the function $F_{WS}(r)$ is reparametrized as:

$$F_{WS}(r) = \left[1 + \exp\left(\frac{r - r_{0,x}A^{1/3}}{a_{0,x}}\right)\right]^{-1} \tag{15}$$

where $x = U, W$.

Table 1. Best-fitting parameters of the 6pWS optical potential. Data for fitting are from Ref. [33, 34]. Antiproton momenta for the given targets are 608 MeV/c (^{12}C and ^{40}Ca) and 305 MeV/c (^{208}Pb). Errors are estimated automatically by ROOT algorithms. The reduced-χ^2 of the fit, defined as $\tilde{\chi}^2 = \chi^2/NDF$, is also reported.

Target	^{12}C	^{40}Ca	^{208}Pb
U_0(MeV)	79.2 ± 0.7	28 ± 2	4.0 ± 0.6
W_0(MeV)	50.8 ± 2	22 ± 3	9 ± 1
$r_{0,U}$(fm)	1.066 ± 0.002	1.34 ± 0.02	1.40 ± 0.04
$r_{0,W}$(fm)	1.26 ± 0.02	1.35 ± 0.02	1.40 ± 0.02
$a_{0,U}$(fm)	0.361 ± 0.004	0.38 ± 0.02	0.30 ± 0.05
$a_{0,W}$(fm)	0.465 ± 0.007	0.55 ± 0.03	0.50 ± 0.02
$\tilde{\chi}^2$	$58.9/42 \approx 1.4$	$95.6/38 \approx 2.5$	$78/13 \approx 6.0$

From Table 1, a decreasing in values of U_0 and W_0 with the increasing of A can be noticed. However, it is very difficult to find a precise relation between the two quantities with such poor analysis. For ^{40}Ca and ^{208}Pb, $r_{0,U} = r_{0,W}$, a common assumption in this kind of model (see for example Ref. [34]); on the contrary, for ^{12}C this is not valid ($r_{0,U} < r_{0,W}$). We also let $a_{0,U}$ and $a_{0,W}$ as free-parameters, obtaining for all nuclei $a_{0,U} > a_{0,W}$. The values of $a_{0,x}$ are quite stable for increasing A, while $r_{0,x}$ slightly increase with A. Concerning the real and imaginary strengths of the potential, with exception of the ^{208}Pb, we observe $U_0 > W_0$ in ^{12}C and ^{40}Ca. The interpretation of these results at the moment is not completely clear and requires a deeper and more complete analysis.
We have also to clarify that important correlation values are found for some parameters pairs.

In Figures 1, 2 and 3, we compare the best fit line - together with the 95% CI band estimation - with the data used for fitting. The agreement is satisfactory for

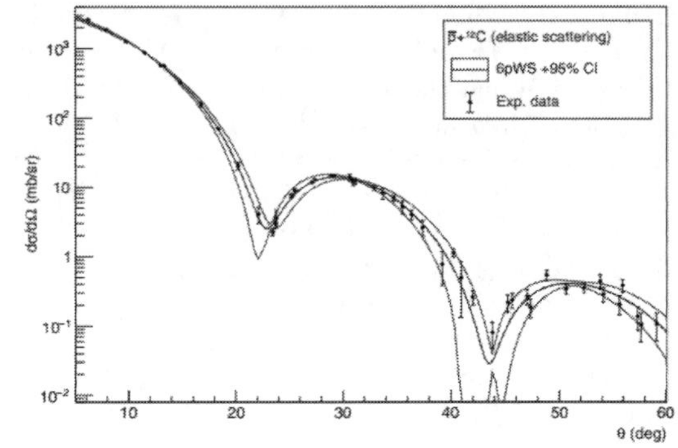

Figure 1. ^{12}C data at p = 608 MeV/c and best fit line with 6pWS with the 95% confidence interval. The reported parameters are from Table 1, while data are from Ref. [33].

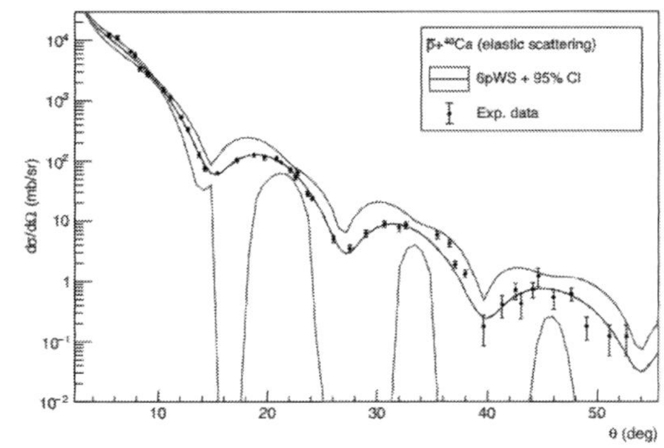

Figure 2. ^{40}Ca data at p = 608 MeV/c and best fit line with 6pWS with the 95% confidence interval. The reported parameters are from Table 1, while data are from Ref. [33].

^{12}C and ^{40}Ca targets. Some discrepancies are present in the ^{208}Pb case at small angles. Furthermore, the fitting procedure gives different problems with the latter target, since the algorithms have difficulty to converge to a unique solution. Therefore, from now on, we will not consider ^{208}Pb in our analysis and we will reconsider it in the future with an improved fitting procedure.

The 6pWS model is then applied to the annihilation process for ^{12}C and ^{40}Ca. We assume that the same parameters determined from elastic scattering process are valid also in this case. The annihilation cross sections in dependence of the antiproton momentum p is calculated using Eqn. 13 with the parameters in Table 1. The results of such calculations for p from 50 MeV/c to 1000 MeV/c can be found in Figures 4 and 5. For the ^{12}C case, we compare the results with data from different measurement: the 100 MeV/c by the ASACUSA collaboration [18] and the measurements by Nakamura *et al.* [35] between 450 MeV/c and 900 MeV/c. In addition, a compar-

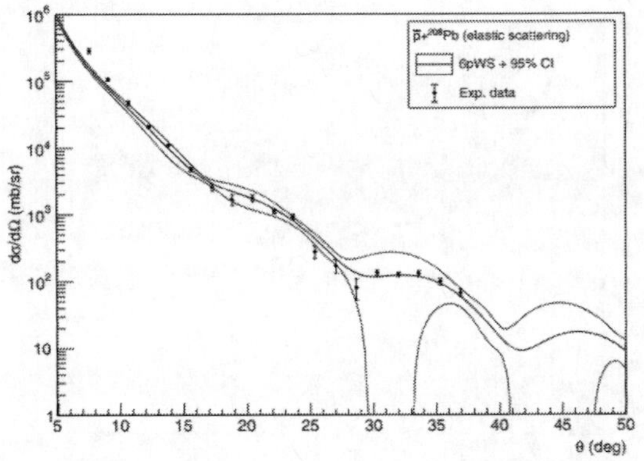

Figure 3. ^{208}Pb data at $p = 305\,\mathrm{MeV/c}$ and best fit line with 6pWS with the 95% confidence interval. The reported parameters are from Table 1, while data are from Ref. [33].

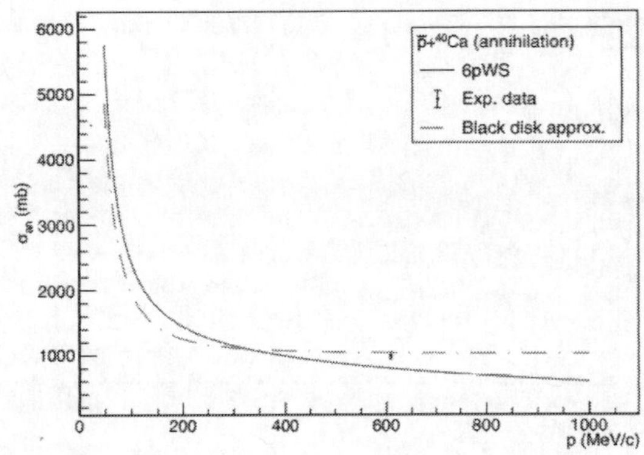

Figure 5. ^{40}Ca annihilation cross section calculation and data. The parameters are from Table 1.

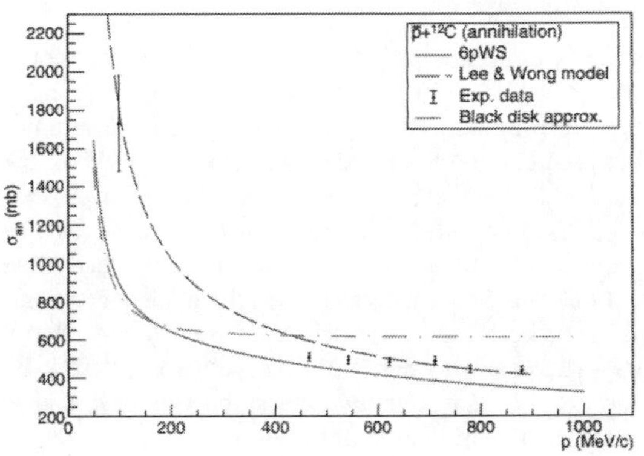

Figure 4. ^{12}C annihilation cross section calculations and data. The parameters are from Table 1.

ison with two different models is done. The blue dashed line is the fitting result of the optical potential proposed by Lee and Wong where they include the dependence of momentum for the real strength $U_0(k)$ [36]. The red dot-dashed line is the "black disk" approximation by Batty *et al.* [37] considering the low-energy regime of the antiproton with the Coulomb focusing due to the charge of the nucleus:

$$\sigma_{an}(k) \approx \pi R^2 \left(1 + \frac{2mZ_t e^2}{\hbar^2 k_{cm} k_{lab} R}\right) \tag{16}$$

where $R = (1.840 + 1.120\,A^{1/3})$ fm and m is the antiproton mass. $k_{cm} = p_{cm}/\hbar$ is the center-of-mass frame wave number[2] and $k_{lab} = p/\hbar$ is the lab-frame wave number.

[2]The center-of-mass momentum p_{cm} can be calculated using the relativistic formula:

$$p_{cm} = \frac{1}{2\sqrt{s}}\left\{\left[s - (m_{\bar{p}} - m_A)^2\right]\left[s - (m_{\bar{p}} + m_A)^2\right]\right\}^{\frac{1}{2}}$$

where $s = (p^{\mu}_{\bar{p}} + p^{\mu}_{A})^2$ (the first Mandelstam variable), and $m_{\bar{p}}$, m_A are the antiproton and nucleus masses.

The 6pWS model results are in agreement with the available carbon-12 data within 2-3 standard-deviation. This is quite satisfying especially considering that our parameters are determined from the fitting to elastic scattering data and not to annihilation data. The behaviour of the curve and the distance from experimental data are quite good, considering the assumptions.

The ^{40}Ca case is more difficult to analyse, due to the scarcity of annihilation cross sections data. The only one shown in Fig. 5 is from Garreta *et al.* [33], at $p = 608\,\mathrm{MeV/c}$. The single datum of ^{40}Ca is one standard-deviation within the black disk approximation and about 4 standard-deviations within our 6pWS prediction.

Both in the ^{12}C and ^{40}Ca analyses, our predictions underestimate the experimental values. Nevertheless, the calculated values are in quite good agreement with available data, reproducing also quite well the behaviour in dependence of the momentum. Therefore, our obtained parameters are enough reliable to use them to make predictions in low-energy regime for differential elastic cross sections.

In Fig.s 6 and 7 for momenta greater than $50\,\mathrm{MeV/c}$ the Coulomb (dashed lines) and total contribution (solid lines) are clearly distinguishable at large angles. On the contrary, for $p < 50\,\mathrm{MeV/c}$ it is difficult to clearly see in a measurement the nuclear contributions from the Rutherford ones. These observations are valid for both ^{12}C and ^{40}Ca targets, as can be seen in the figures mentioned above. The 6pWS model, with the parameters obtained by the fitting, therefore suggests the possibility to see nuclear effects in elastic scattering measurement for $p > 50\,\mathrm{MeV/c}$ and at large angles. Under this momentum (and for angles smaller than 90°), it seems to be very difficult to measure such contributions. Similar results for the angular distribution at low energy were found by Protasov *et al.* [38] for the ^{12}C case using a different parametrization of the optical potential.

As shown in Fig. 8, in their case (solid blue line) nuclear effects at 50 MeV/c are more accentuated than in our

https://doi.org/10.1051/epjconf/202226201018

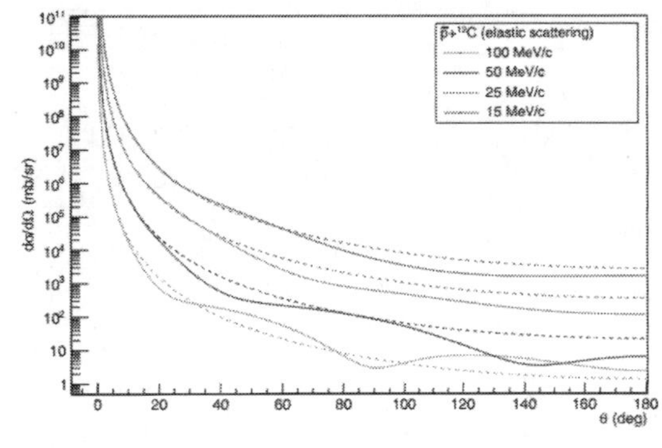

Figure 6. ^{12}C differential elastic cross section in low momentum regime, from 15 MeV/c to 100 MeV/c. In dashed line the correspondent Coulomb cross sections is shown. The parameters are taken from Table 1.

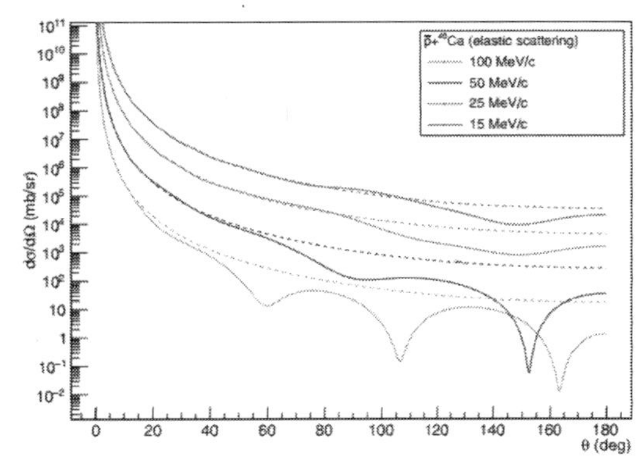

Figure 7. ^{40}Ca differential elastic cross section in low momentum regime, from 15 MeV/c to 100 MeV/c. In dashed line the correspondent Coulomb cross sections is shown. The parameters are taken from Table 1.

calculations (solid black line), resulting in a lower limit estimation for measurements ($p > 20$ MeV/c). This discrepancy between our model and theirs is complicate to be understood. One possible explanation could be the different parametrization of the potential and the different kind of parameters used. However, both the behaviour of the curve and the position of the minimum around 145 MeV/c are very similar in the two predictions.

4 Summary and future developments

A six free-parameter Woods-Saxon (6pWS) optical potential is fitted to differential elastic scattering cross section data available for antiproton projectiles on medium-light (^{12}C, ^{40}Ca) and heavy (^{208}Pb) nuclei at intermediate energy. We validate these results comparing the annihilation cross section measurements with the calculations of our model, confirming a satisfying agreement with available data. The agreement is quite good (2-4 experimental errors), considering the free-parameters of the model are obtained from different types of data, some discrepancies are present. Then, we use the obtained parameters to predict the differential elastic cross sections at low momenta ($p < 100$ MeV/c), focusing on large-angles behaviour. Nuclear effects seems to be observable for momentum greater than 50 MeV/c, while under this value Coulomb contribution begins to hide the nuclear contribution. In our knowledge, no such predictions at low energy exists in literature (a part from the very recent work by Protasov *et al.* [38]) and also measurements of this kind do not exist, in particular at large angles.

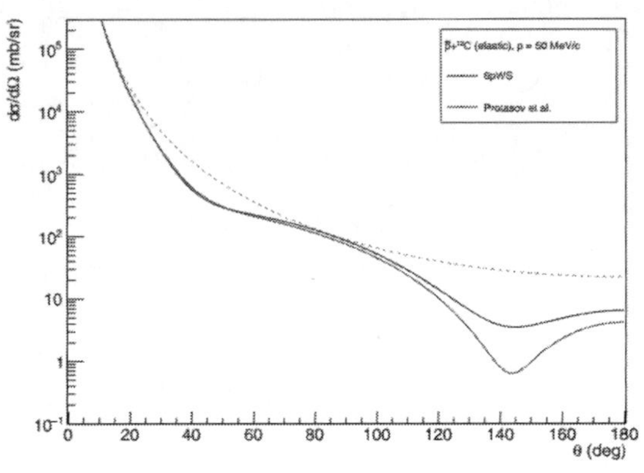

Figure 8. ^{12}C differential elastic cross section at $p = 50$ MeV/c. In solid black line, our model; in solid blue line the calculation by Protasov *et al.* [39] The dashed green line is the Rutherford scattering cross section.

Some improvements are necessary to our model. A Lee and Wong approach, introducing dependences on antiproton momentum, is a possible path to follow. Moreover, some other terms can be introduced in the potential, *e.g.* an absorption surface term proportional to $dF_{WS}(r)/dr$. Also, we have to consider other possible parametrizations of the optical potential more suitable for our purpose, like the one proposed by Batty, Friedman and Gal [40]. Other future developments of this analysis can include better evaluation of uncertainties and correlations of the parameters, and the use of more efficient algorithms. These improvements on antiproton analysis are useful to design antiproton experiments at low energy, *e.g.* at ELENA (Extra Low ENergy Antiproton) ring [41] at AD and at FLAIR (Facility for Low-energy Antiproton and Ion Research) at GSI [42, 43]. Finally, a fundamental step will be to include the present calculations and analysis for antineutrons. The extension of our analysis to antineutrons scattering can be useful to dispel some incongruities highlighted by Friedman [44]. Moreover, the determination of the scattering length of the antineutron is a necessary step to design experiments (*e.g.*, for neutron-antineutron oscillation [38, 39, 45]). Further measurements of annihilation – in particular between 100 and 450 MeV/c – and nuclear elastic scattering cross sections at low energies ($p > 50$ MeV/c) and large angles ($\theta > 90°$)

could be useful to improve phenomenological models for nuclear strong interactions. Facilities which could provide in future such experimental conditions are ELENA and FLAIR.

References

[1] R. Bizzarri, P. Guidoni, F. Marcelja, F. Marzano, E. Castelli, M. Sessa, Il Nuovo Cimento A (1965-1970) **22**, 225 (1974)

[2] B. Gunderson, J. Learned, J. Mapp, D. Reeder, Physical Review D **23**, 587 (1981)

[3] F. Balestra, Y.A. Batusov, G. Bendiscioli, M. Bussa, L. Busso, I. Falomkin, L. Ferrero, V. Filippini, G. Fumagalli, G. Gervino et al., Physics Letters B **149**, 69 (1984)

[4] F. Balestra, S. Bossolasco, M.P. Bussa, L. Bussa, L. Ferrero, D. Panzieri, G. Piragino, F. Tosello, C. Guaraldo, A. Maggiora et al., Physics Letters B **165**, 265 (1985)

[5] M. Agnello, F. Iazzi, B. Minetti, L. Cugusi, M. Macciotta, S. Marcello, A. Masoni, G. Puddu, A. Raimondi, S. Serci et al., EPL (Europhysics Letters) **7**, 13 (1988)

[6] F. Balestra, R. Barbieri, Y.A. Batusov, G. Bendiscioli, S. Bossolasco, F. Breivik, M.P. Bussa, L. Busso, C. Guaraldo, I. Falomkin et al., Physics Letters B **230**, 36 (1989)

[7] W. Brückner, B. Cujec, H. Döbbeling, K. Dworschak, F. Güttner, H. Kneis, S. Majewski, M. Nomachi, S. Paul, B. Povh et al., Zeitschrift für Physik A Atomic Nuclei **335**, 217 (1990)

[8] V. Ableev, A. Adamo, M. Agnello, F. Balestra, G. Belli, G. Bendiscioli, A. Bertin, P. Boccaccio, G. Bonazzola, E. Botta et al., Il Nuovo Cimento A (1965-1970) **107**, 943 (1994)

[9] C. Barbina, A. Ahmidouch, R. Birsa, F. Bradamante, A. Bressan, S. Dalla Torre-Colautti, M. Giorgi, R. Hess, F. Iazzi, R. Kunne et al., Nuclear Physics A **612**, 346 (1997)

[10] A. Benedettini, A. Bertin, M. Bruschi, M. Capponi, A. Collamati, I. D'Antone, S. De Castro, R. Dona, A. Ferretti, D. Galli et al., Nuclear Physics B-Proceedings Supplements **56**, 58 (1997)

[11] A. Zenoni, A. Bianconi, F. Bocci, G. Bonomi, M. Corradini, A. Donzella, E.L. Rizzini, L. Venturelli, A. Bertin, M. Bruschi et al., Physics Letters B **461**, 405 (1999)

[12] A. Zenoni, A. Bianconi, G. Bonomi, M. Corradini, A. Donzella, E.L. Rizzini, L. Venturelli, A. Bertin, M. Bruschi, M. Capponi et al., Physics Letters B **461**, 413 (1999)

[13] A. Bianconi, G. Bonomi, M. Bussa, E.L. Rizzini, L. Venturelli, A. Zenoni, G. Pontecorvo, C. Guaraldo, F. Balestra, L. Busso et al., Physics Letters B **481**, 194 (2000)

[14] A. Bianconi, G. Bonomi, M. Bussa, E.L. Rizzini, L. Venturelli, A. Zenoni, G. Pontecorvo, C. Guaraldo, F. Balestra, L. Busso et al., Physics Letters B **492**, 254 (2000)

[15] F. Iazzi, A. Feliciello, M. Agnello, M. Astrua, E. Botta, T. Bressani, D. Calvo, S. Costa, F. D'Isep, A. Filippi et al., Physics Letters B **475**, 378 (2000)

[16] M. Astrua, E. Botta, T. Bressani, D. Calvo, C. Casalegno, A. Feliciello, A. Filippi, S. Marcello, M. Agnello, F. Iazzi, Nuclear Physics A **697**, 209 (2002)

[17] A. Bianconi, M. Corradini, M. Hori, M. Leali, E.L. Rizzini, V. Mascagna, A. Mozzanica, M. Prest, E. Vallazza, L. Venturelli et al., Physics Letters B **704**, 461 (2011)

[18] H. Aghai-Khozani, A. Bianconi, M. Corradini, R. Hayano, M. Hori, M. Leali, E. Lodi Rizzini, V. Mascagna, Y. Murakami, M. Prest et al., Nuclear Physics A **970**, 366 (2018)

[19] M. Corradini, R. Hayano, M. Hori, M. Leali, E. Lodi Rizzini, V. Mascagna, A. Mozzanica, M. Prest, K. Todoroki, E. Vallazza et al., Nuclear Instruments and Methods in Physics Research Section A: Accelerators, Spectrometers, Detectors and Associated Equipment **711**, 12 (2013)

[20] K. Todoroki, D. Barna, R. Hayano, H. Aghai-Khozani, A. SÃ³tÃ©r, M. Corradini, M. Leali, E. Lodi-Rizzini, V. Mascagna, L. Venturelli et al., Nuclear Instruments and Methods in Physics Research Section A: Accelerators, Spectrometers, Detectors and Associated Equipment **835**, 110 (2016)

[21] M. Corradini, M. Leali, E.L. Rizzini, V. Mascagna, M. Prest, E. Vallazza, L. Venturelli, Hyperfine Interactions **233**, 53 (2015)

[22] H. Aghai-Khozani, D. Barna, M. Corradini, D. De Salvador, R. Hayano, M. Hori, M. Leali, E. Lodi-Rizzini, V. Mascagna, M. Prest et al., Nuclear Physics A **1009**, 122170 (2021)

[23] A.G. Cohen, A.D. Rújula, S.L. Glashow, **495**, 539 (1998)

[24] H. Kurki-Suonio, E. Sihvola, Phys. Rev. Lett. **84**, 3756 (2000)

[25] K. Abe, H. Fuke, S. Haino, T. Hams, A. Itazaki, K. Kim, T. Kumazawa, M. Lee, Y. Makida, S. Matsuda et al., Physics Letters B **670**, 103 (2008)

[26] T. Aramaki, S. Boggs, P. von Doetinchem, H. Fuke, C. Hailey, S. Mognet, R. Ong, K. Perez, J. Zweerink, Astroparticle Physics **59**, 12 (2014)

[27] L. Gray, T.E. Kalogeropoulos, Radiation Research **97**, 246 (1984)

[28] M.H. Holzscheiter, N. Bassler, N. Agazaryan, G. Beyer, E. Blackmore, J.J. DeMarco, M. Doser, R.E. Durand, O. Hartley, K.S. Iwamoto et al., Radiotherapy and Oncology **81**, 233 (2006)

[29] N. Bassler, M.H. Holzscheiter, H. Knudsen, AIP Conference Proceedings **796**, 423 (2005)

[30] H. Feshbach, Annual Review of Nuclear Science **8**, 49 (1958)

[31] N.S. Chant, P.G. Roos (1998), (unpublished)

[32] R. Brun, F. Rademakers, P. Canal, A. Naumann, O. Couet, L. Moneta, V. Vassilev, S. Linev, D. Piparo, G. GANIS et al., *root-project/root: v6.18/02* (2019), `https://doi.org/10.5281/zenodo.3895860`

[33] D. Garreta, P. Birien, G. Bruge, A. Chaumeaux, D. Drake, S. Janouin, D. Legrand, M. Lemaire, B. Mayer, J. Pain et al., Physics Letters B **149**, 64 (1984)

[34] S. Janouin, M.C. Lemaire, D. Garreta, P. Birien, G. Bruge, D. Drake, D. Legrand, B. Mayer, J. Pain, J. Peng et al., Nuclear Physics A **451**, 541 (1986)

[35] K. Nakamura, J. Chiba, T. Fujii, H. Iwasaki, T. Kageyama, S. Kuribayashi, T. Sumiyoshi, T. Takeda, H. Ikeda, Y. Takada, Phys. Rev. Lett. **52**, 731 (1984)

[36] T.G. Lee, C.Y. Wong, Phys. Rev. C **97**, 054617 (2018)

[37] C. Batty, E. Friedman, A. Gal, Nuclear Physics A **689**, 721 (2001)

[38] K.V. Protasov, V. Gudkov, E.A. Kupriyanova, V.V. Nesvizhevsky, W.M. Snow, A.Y. Voronin, Phys. Rev. D **102**, 075025 (2020)

[39] V.V. Nesvizhevsky, V. Gudkov, K.V. Protasov, W.M. Snow, A.Y. Voronin, Phys. Rev. Lett. **122**, 221802 (2019)

[40] C. Batty, E. Friedman, A. Gal, Physics Reports **287**, 385 (1997)

[41] V. Chohan, C. Alanzeau, M. Angoletta, J. Baillie, D. Barna, W. Bartmann, P. Belochitskii, J. Borburgh, H. Breuker, F. Butin et al., *Extra Low ENergy Antiproton (ELENA) ring and its Transfer Lines: Design Report* (2014), ISBN 9789290834007

[42] E. Widmann, Physica Scripta **2015**, 014074 (2015)

[43] C.P. Welsch, A. Papash, O. Gorda, J. Harasimowicz, O. Karamyshev, G. Karamysheva, D. Newton, M. Panniello, M. Putignano, M. Siggel-King et al., Hyperfine Interactions **213**, 205 (2012)

[44] E. Friedman, Nuclear Physics A **925**, 141 (2014)

[45] J.L. Barrow, A.S. Botvina, E.S. Golubeva, J.M. Richard, *A new model of intranuclear neutron-antineutron transformations in* $^{16}_{8}o$ (2021), `arXiv:2111.10478[hep-ex]`

EPJ Web of Conferences **262**, 01019 (2022)
EXA2021
https://doi.org/10.1051/epjconf/202226201019

Role of the effective range in the weak-binding relation

Tomona Kinugawa[1,*] and *Tetsuo* Hyodo[1,**]

[1]Department of Physics, Tokyo Metropolitan University, Hachioji 192-0397, Japan

Abstract. We study the range correction in the weak-binding relation, which relates the internal structure of hadrons with the scattering length and the binding energy. Utilizing the effective field theories, we show that the effective range originates from the derivative coupling interaction as well as from the channel coupling to the bare state, and that the different contributions are not distinguishable. By examining the compositeness in the effective field theories, it is demonstrated that the effective range induces the finite range correction for the weak-binding relation in addition to the previously known contributions. We thus propose to include the range correction in the uncertainty terms of the weak-binding relation.

1 Introduction

The number of the observed hadrons has been continuously increasing [1], and there are several candidates which seem to require exotic configurations beyond the simple $\bar{q}q$ or qqq structures, such as multiquarks, hadronic molecules, hybrid states, and so on [2, 3]. The investigation of the exotic structures of hadrons is thus one of the major subjects in hadron physics. In particular, the recent observation of the doubly charmed tetraquark T_{cc} by the LHCb collaboration [4, 5] further stimulates this trend.

Although there are many approaches to unveil the internal structure of hadrons, it is desirable to relate the hadron structure with observable quantities in a model-independent manner. One promising method is the weak-binding relation for bound states by Weinberg [6]

$$a_0 = R\left(\frac{2X}{1+X} + O\left(\frac{R_{\rm typ}}{R}\right)\right), \tag{1}$$

where the compositeness X, the fraction of the hadronic molecule component, is related with the scattering length a_0 and the length scale $R = 1/\sqrt{2\mu B}$ determined by the binding energy B. The higher order correction terms are of the order of $O(R_{\rm typ}/R)$ with the typical length scale of the microscopic two-body interactions $R_{\rm typ}$. When the binding energy B is sufficiently small such that $R_{\rm typ}/R \ll 1$, then the compositeness X can be estimated by the observables, a_0 and B. This means that the internal structure of a bound state can be studied, even without having detailed knowledge of the wavefunction of the state. This feature should be useful in hadron physics, because the experimental data allow us to determine only a few scattering parameters at best. Indeed, the notion of the compositeness as well as the generalization of the weak-binding relation have been recently discussed in the context of hadron physics [7–13].

*e-mail: kinugawa-tomona@ed.tmu.ac.jp
**e-mail: hyodo@tmu.ac.jp

The weak-binding relation is closely related to the low-energy universality for the systems with a large scattering length [14, 15]. It is known that, when the system has a shallow bound state, the scattering length can be much larger than the interaction scale, and the microscopic details of the system becomes irrelevant for the low-energy physics. It follows in the strict weak-binding limit ($B \to 0$) that any lengths in the system should scale with the scattering length, which implies that $a_0 = R$. The weak-binding relation (1) tells us how the ideal relation $a_0 = R$ is modified by the coupling to the other components ($2X/(1+X)$ with $X < 1$) and by the finite range effect ($O(R_{\rm typ}/R)$). In the previous studies, the length scale $R_{\rm typ}$ was estimated by the interaction range, which reflects the finite range correction from the off-shell nature of the system. On the other hand, there exists yet another length scale, the effective range, which reflects the range correction in the on-shell scattering amplitude.

Here we would like to focus on the finite range correction in the weak-binding relation (1), stemming from the effective range [16]. To this end, we examine the origin of the effective range in several effective field theories. With the analysis of the compositeness of the bound state, we then propose a method to include the range correction in the weak-binding relation.

2 Effective field theories

Effective field theories are useful tools to describe the low-energy behaviors of given physical systems. We summarize properties of the nonrelativistic effective field theories for the low-energy two-body scattering following Ref. [17], and show that the effective range can be induced either by the derivative coupling or the inclusion of the additional bare field.

2.1 Zero-range model

The simplest effective field theory contains the four-point contact interaction, called the zero-range model, whose Hamiltonian is given by

$$\mathcal{H}_{\rm ZR} = \frac{1}{2m}\nabla\psi^\dagger \cdot \nabla\psi + \frac{\lambda_0}{4}(\psi^\dagger\psi)^2, \tag{2}$$

where ψ is the boson field with mass m, and λ_0 is the bare coupling constant. The four-point contact interaction corresponds to the delta function potential in the coordinate space. Because of the singularity at the origin which is stronger than the kinetic term, the delta function potential in the three-dimensional space should be regularized to avoid the ultraviolet divergence. In the effective field theory, the momentum integration should be cut off at some ultraviolet scale Λ. By regarding Eq. (2) as an effective description of the underlying microscopic theory with a finite range interaction, we can interpret Λ as the momentum scale below which the interaction can be well approximated by the pointlike one, as in Eq. (2). In other words, the inherent interaction range in the microscopic theory should be estimated by $\sim 1/\Lambda$.

It is known that the field theory (2) is renormalizable in the two-body sector. Namely, one can absorb the cutoff Λ dependence by the bare parameter λ_0, without modifying the observables. By taking the $\Lambda \to \infty$ limit after the renormalization, one finds that the two-body scattering amplitude in the zero-range model is determined exclusively by the scattering length a_0:

$$f_{\rm ZR}(k) = \left(-\frac{1}{a_0} - ik\right)^{-1}. \tag{3}$$

Because we interpret $1/\Lambda$ as the interaction range, the $\Lambda \to \infty$ limit corresponds to the zero range limit where the interaction range is formally set to zero. This model captures the basic aspects of the low-energy universality. In fact, Eq. (3) has a pole at $k = i/a_0$, which represents the bound state for a positive a_0. Noting that $R = i/k$, one finds the relation

$$a_0 = R, \tag{4}$$

as indicated by the low-energy universality.

2.2 Effective range model

The first correction to the zero-range model can be obtained by introducing a derivative coupling term. This is called the effective range model:

$$\mathcal{H}_{\mathrm{ER}} = \frac{1}{2m}\nabla\psi^\dagger \cdot \nabla\psi + \frac{\lambda_0}{4}(\psi^\dagger\psi)^2 + \frac{\rho_0}{4}\nabla(\psi^\dagger\psi)\cdot\nabla(\psi^\dagger\psi), \tag{5}$$

where ρ_0 is the coupling constant in the derivative coupling interaction. Performing the same renormalization procedure with the zero-range model, we end up with the scattering amplitude in the $\Lambda \to \infty$ limit:

$$f_{\mathrm{ER}}(k) = \left(-\frac{1}{a_0} + \frac{r_e}{2}k^2 - ik\right)^{-1}, \tag{6}$$

which now contains the effective range r_e, because the Hamiltonian (5) consists of two coupling constants λ_0 and ρ_0, and there is a sensible $\Lambda \to \infty$ limit where a_0 and r_e remain finite. Note however that the interaction range is zero in the $\Lambda \to 0$ limit, and therefore, the effective range r_e is induced by the derivative coupling term. We also note that the effective range in Eq. (6) is always negative. This is consistent with the Wigner causality bound [18, 19], which imposes the upper bound of the effective range to be of the order of the interaction range. On the other hand, with a finite but sufficiently large Λ, r_e can be positive below the Wigner bound $12/(\pi a_0^3\Lambda^3)(a_0\Lambda - \pi/2)^2$ [20].

Because the denominator of Eq. (6) is quadratic in k, the amplitude always contains two poles. Substituting $R = i/k$ for the bound state condition $1/f_{\mathrm{ER}}(k) = 0$ and picking up the solution which is closer to the $k = 0$, we obtain the relation between a_0, R, and r_e as

$$a_0 = R\frac{2r_e/R}{1-(r_e/R-1)^2} = R\frac{2}{2-r_e/R}. \tag{7}$$

This reduces to $a_0 = R$ for the vanishing effective range $r_e \to 0$. For finite r_e, however, a_0 deviates from R. Because $R > 0$, the negative effective range implies $a_0 < R$.

2.3 Resonance model

It is also possible to introduce an additional bare field ϕ which couples to the two-body scattering state,

$$\mathcal{H}_{\mathrm{R}} = \frac{1}{2m}\nabla\psi^\dagger\cdot\nabla\psi + \frac{1}{4m}\nabla\phi^\dagger\cdot\nabla\phi + \nu_0\phi^\dagger\phi + \frac{g_0}{2}(\phi^\dagger\psi^2 + \psi^{\dagger 2}\phi) + \frac{\lambda_0}{4}(\psi^\dagger\psi)^2, \tag{8}$$

where ν_0 is the energy of the bare state measured from the threshold of the scattering state and g_0 is the coupling strength between the scattering state and the bare field. The scattering amplitude in the zero range limit $\Lambda \to \infty$ is obtained as

$$f_{\mathrm{R}}(k) = \left(-\frac{1}{a_0} + \frac{r_e}{2}k^2 + O(k^4) - ik\right)^{-1}, \tag{9}$$

where r_e and the higher order terms in the effective range expansion also appear in the amplitude, because the elimination of the bare field ϕ induces the energy dependence in the effective two-body interaction. Again in this case, the interaction range is set to zero, so the effective range originates in the coupling to the bare field. Also, the effective range is always negative, in accordance with the Wigner bound. In contrast to the effective range model, in the resonance model, r_e is always negative even with a finite Λ. To describe the system with a positive effective range, one needs to introduce the ghost field with negative norm [17, 21].

In this model, the expression of R cannot be written down in a closed form by the observables, because of the higher order terms in the effective range expansion. It is however clear that the $a_0 = R$ relation does not hold in general. The expression of R reduces to Eq. (7) when the higher order terms are neglected.

3 Weak-binding relation and range correction

Now we consider the structure of the bound state in the above field theories from the viewpoint of the weak-binding relation and the compositeness.

3.1 Compositeness and the weak-binding relation

As shown in Refs. [11, 13] the Weinberg's weak-binding relation can be derived in the effective field theories. The compositeness X is defined as the overlap integral of the bound state $|\,B\,\rangle$, which is the eigenstate of the full Hamiltonian, and the two-body scattering states $|\,\boldsymbol{p}\,\rangle$, the eigenstates of the free Hamiltonian:

$$X = \int \frac{d^3\boldsymbol{p}}{(2\pi)^3} |\langle\,\boldsymbol{p}\,|\,B\,\rangle|^2, \tag{10}$$

which is the probability of finding the two-body components in the bound state wavefunction.

We first consider the value of X from the completeness relation in the effective field theories. It can be shown by the phase symmetry of the ψ field in the Hamiltonians (2) and (5) that the particle number is conserved. This means that the completeness relation spanned by the eigenstates of the free Hamiltonian in the zero-range model and the effective range model for the two-body sector is given only by the scattering states:

$$1_{\rm ZR} = 1_{\rm ER} = \int \frac{d^3\boldsymbol{p}}{(2\pi)^3} |\,\boldsymbol{p}\,\rangle\langle\,\boldsymbol{p}\,|. \tag{11}$$

Using the normalization condition for the bound state $\langle\,B\,|\,B\,\rangle = 1$, we obtain the value of the compositeness

$$X_{\rm ZR} = X_{\rm ER} = 1. \tag{12}$$

In other words, by definition, the exact value of the compositeness is unity in the zero-range and effective range models. On the other hand, from the phase symmetry, the completeness relation in the resonance model is given by the sum of the bare state ϕ and the scattering states [11, 13]:

$$1_{\rm R} = |\,\phi\,\rangle\langle\,\phi\,| + \int \frac{d^3\boldsymbol{p}}{(2\pi)^3} |\,\boldsymbol{p}\,\rangle\langle\,\boldsymbol{p}\,|. \tag{13}$$

In this case, we obtain

$$X_{\rm R} < 1, \tag{14}$$

Table 1. Summary of the effective range r_e, the exact value of the compositeness X_{exact}, and X_{WB} determined by the weak-binding relation (1) in the effective field theories in the zero range limit.

EFT	r_e	X_{exact}	X_{WB}
Zero-range model	$r_e = 0$	$X = 1$	$X = 1$
Effective range model	$r_e < 0$	$X = 1$	$X < 1$
Resonance model	$r_e < 0$	$X < 1$	$X < 1$

because $\langle B|\phi\rangle\langle\phi|B\rangle = |\langle\phi|B\rangle|^2 > 0$ provided that $g_0 \neq 0$.

Next, we use the weak-binding relation (1) to determine the compositeness X in these theories. Recalling that the interaction range is related to the cutoff as $R_{\text{typ}} = 1/\Lambda$, in the zero range limit $\Lambda \to \infty$, the correction terms $O(R_{\text{typ}}/R)$ vanish, and Eq. (1) determines X only by a_0 and R. For $a_0 = R$ in the zero-range model, we obtain $X_{\text{ZR}} = 1$, which coincides with the exact value in Eq. (12). On the other hand, in the effective range model, we obtain $a_0 < R$, which corresponds to $X_{\text{ER}} < 1$. This is in contradiction to the exact value (12). For the resonance model, the weak-binding relation gives $X_{\text{R}} < 1$, in accordance with the exact value (14).

The above results are summarized in Table 1. It is clear that the weak-binding relation (1) with $R_{\text{typ}} = 1/\Lambda$ works for the zero-range model and the resonance model, but in the effective range model, the weak-binding relation does not reproduce the exact value. The discrepancy cannot be attributed to the correction terms which vanish in the zero range limit. The origin of the discrepancy is traced back to the finite effective range which violates the universal relation $a_0 = R$. At the same time, in the resonance model, the finite effective range does not provide contradiction in the weak-binding relation, because the exact value of the compositeness is modified by the bare state contribution.

The models in Table 1 have a negative effective range r_e, which correspond to, for instance, the $D\bar{D}^*$ scattering coupled with the exotic hadron $X(3872)$ [22]. On the other hand, most of hadron scatterings have a positive r_e such as the pn scattering coupled with the deuteron [23]. As we mention above, it is possible to realize the positive r_e case by using the finite cutoff Λ in the effective range model. However, the finite $\Lambda = 1/R_{\text{typ}}$ induces the finite correction terms in weak-binding relation (1). Because the estimated compositeness X_{WB} depends on the correction terms, we cannot compare X_{WB} with $X_{\text{exact}} = 1$ without numerical calculations in the positive r_e case. A detailed quantitative analysis will be shown in Ref. [20].

3.2 Origin of the effective range

We have shown that the effective range r_e originates either from the derivative coupling interaction (effective range model) or from the channel coupling to the bare state (resonance model). While the former does not modify the value of the compositeness, the latter induces $X_{\text{R}} < 1$ because of the bare state contribution. In other words, the origin of the effective range is two-fold. In order to determine the compositeness from the observables, one may wonder if we can distinguish the origin of the effective range.

For this purpose, we consider the effective field theory which contains both the derivative coupling interaction in Eq. (5) and the coupling to the bare field in Eq. (8). Calculating the effective range with a finite cutoff Λ, we obtain

$$r_e = \frac{16\pi}{m}\frac{\left[1 + \frac{m}{12\pi^2}\rho_0\Lambda^3\right]^2\left\{2\rho_0\left[1 + \frac{m}{24\pi^2}\rho_0\Lambda^3\right] - \frac{g_0^2}{m\nu_0^2}\right\}}{\left[\lambda_0 - \frac{m}{20\pi^2}\rho_0^2\Lambda^5 - \frac{g_0^2}{\nu_0}\right]^2}. \tag{15}$$

In this expression, the derivative coupling contribution is in the terms with ρ_0, and the bare state contribution is in the g_0^2 term. By setting $\rho_0 = 0$ ($g_0 = 0$), Eq. (15) reduces to the expression of the effective range (resonance) model. Because the terms with ρ_0 and those with g_0^2 are entangled in Eq. (15), we conclude that the origin of the effective range is not distinguishable.

3.3 Range correction

The above conclusion indicates that the discrepancy between X_{exact} and X_{WB} in the effective range model cannot be resolved by the correction of the $2X/(1+X)$ term with r_e, because its origin is not determined. Rather, the range correction should be included in the correction term $O(R_{\text{typ}}/R)$, by modifying the definition of R_{typ}. Namely, there exists a new correction term, which does not vanish in the zero range limit, $\Lambda \to \infty$.

In the weak-binding situation with a large R, the scattering length in the effective range model (7) can be given by

$$a_0 = R\left\{1 + O\left(\left|\frac{r_e}{R}\right|\right)\right\}. \tag{16}$$

This suggests to redefine the length scale R_{typ} as

$$R_{\text{typ}} = \max\{R_{\text{int}}, R_{\text{eff}}\}, \tag{17}$$

with $R_{\text{int}} = 1/\Lambda$, the range of the interaction in the microscopic theory (off-shell length scale). In Eq. (17), we propose to add R_{eff}, which is the length scale in the effective range expansion, such as $|r_e|$ or the properly normalized coefficients of the higher order terms (on-shell length scale). With this range correction, the discrepancy in the effective range model is resolved, because there remain the correction terms $O(R_{\text{eff}}/R)$ even in the zero range limit $\Lambda \to \infty$ ($R_{\text{int}} \to 0$).

The new definition (17) reduces to the previous one when $R_{\text{eff}} < R_{\text{int}}$. In usual cases without fine tuning, the magnitude of the effective range is of the same order with the interaction range R_{int}, where the R_{eff} term in Eq. (17) would not have a substantial effect. However, because there is no lower limit on the value of r_e, its magnitude can in principle be much larger than the interaction range such that $R_{\text{eff}} \gg R_{\text{int}}$. In fact, it is numerically demonstrated that the improved weak-binding relation with Eq. (17) has larger applicability than the previous one [16]. The effective range model in the zero range limit is one of such examples. Thus, the improvement in Eq. (17) will be important for the systems with a sizable magnitude of the effective range.

4 Summary

We have discussed the role of the effective range in the weak-binding relation. The properties of the low-energy effective field theories are reviewed, with the emphasis on the effective range and its origin. It is shown that the effective range induced by the derivative coupling interaction does not affect the compositeness. This causes the discrepancy between the exact value of the compositeness and that given by the weak-binding relation in the effective range model in the zero range limit. We then propose the range correction of the weak-binding relation by modifying the correction terms, such that the above discrepancy is resolved, and the applicability of the weak-binding relation is enlarged.

The improved relation can be applied to various hadron systems. For instance, the analysis of the T_{cc} by the LHCb collaboration [5] reported the upper bound of the magnitude of the effective range to be 11 fm. Also, a recent analysis of $X(3872)$ suggests that $r_e \sim -5$ fm [22] (see, however, Ref. [24] on the uncertainty of the determination of the effective range in these analysis). These results imply the larger magnitude of the effective range than the hadron interaction scale of 1 fm and the range correction of the weak-binding relation should be important. These systems are currently under investigation, and the results will be reported elsewhere in future.

References

[1] P.A. Zyla et al. (Particle Data Group), PTEP **2020**, 083C01 (2020)
[2] A. Hosaka, T. Iijima, K. Miyabayashi, Y. Sakai, S. Yasui, PTEP **2016**, 062C01 (2016), arXiv:1603.09229
[3] F.K. Guo, C. Hanhart, U.G. Meissner, Q. Wang, Q. Zhao, B.S. Zou, Rev. Mod. Phys. **90**, 015004 (2018), arXiv:1705.00141
[4] R. Aaij et al. (LHCb) (2021), arXiv:2109.01038
[5] R. Aaij et al. (LHCb) (2021), arXiv:2109.01056
[6] S. Weinberg, Phys. Rev. **137**, B672 (1965)
[7] V. Baru, J. Haidenbauer, C. Hanhart, Y. Kalashnikova, A.E. Kudryavtsev, Phys. Lett. B **586**, 53 (2004), hep-ph/0308129
[8] T. Hyodo, D. Jido, A. Hosaka, Phys. Rev. C **85**, 015201 (2012), 1108.5524
[9] F. Aceti, E. Oset, Phys. Rev. D **86**, 014012 (2012), arXiv:1202.4607
[10] T. Sekihara, T. Hyodo, D. Jido, PTEP **2015**, 063D04 (2015), arXiv:1411.2308
[11] Y. Kamiya, T. Hyodo, Phys. Rev. C **93**, 035203 (2016), arXiv:1509.00146
[12] Z.H. Guo, J.A. Oller, Phys. Rev. D **93**, 096001 (2016), arXiv:1508.06400
[13] Y. Kamiya, T. Hyodo, PTEP **2017**, 023 (2017), arXiv:1607.01899
[14] E. Braaten, H.W. Hammer, Phys. Rept. **428**, 259 (2006), cond-mat/0410417
[15] P. Naidon, S. Endo, Rept. Prog. Phys. **80**, 056001 (2017), 1610.09805
[16] T. Kinugawa, T. Hyodo, (2021), arXiv:2111.06619
[17] E. Braaten, M. Kusunoki, D. Zhang, Annals Phys. **323**, 1770 (2008), arXiv:0709.0499
[18] A. Bohm, *Quantum Mechanics: Foundations and Applications, 3rd edition* (Springer, 2001)
[19] I. Matuschek, V. Baru, F.K. Guo, C. Hanhart, Eur. Phys. J. A **57**, 101 (2021), arXiv:2007.05329
[20] T. Kinugawa, T. Hyodo, *in preparation*
[21] D.B. Kaplan, Nucl. Phys. B **494**, 471 (1997), nucl-th/9610052
[22] A. Esposito, L. Maiani, A. Pilloni, A.D. Polosa, V. Riquer (2021), arXiv:2108.11413
[23] R. Machleidt, Phys. Rev. C **63**, 024001 (2001), nucl-th/0006014
[24] V. Baru, X.K. Dong, M.L. Du, A. Filin, F.K. Guo, C. Hanhart, A. Nefediev, J. Nieves, Q. Wang (2021), arXiv:2110.07484

Elucidating the ρ-meson's role as intermediate resonance in the time-like electromagnetic pion form factor

Reinhard Alkofer[1,*], *Ángel S.* Miramontes[2,**], and *Hèlios* Sanchis Alepuz[3,***]

[1]Institute of Physics, University of Graz, NAWI Graz, Universitätsplatz 5, 8010 Graz, Austria
[2]Instituto de Física y Matemáticas, Universidad Michoacana de San Nicolás de Hidalgo, Morelia, Michoacán 58040, Mexico
[3]Silicon Austria Labs GmbH, Inffeldgasse 25F, 8010 Graz, Austria

Abstract. Motivated by the planned measurements of the time-like electromagnetic proton form factors an exploratory study of the time-like electromagnetic pion form factor is presented in a formalism which describes mesons as Poincaré-invariant bound states. In the respective quark interaction kernel, beyond the gluon-intermediated interactions for valence-type quarks, non-valence effects are included by allowing the pions to couple back in a self-consistent manner to the quarks. Consequently, the opening of the dominant ρ decay channel, $\rho \to \pi\pi$, and the presence of a multi-particle branch cut, setting in when the two-pion threshold is crossed, are included consistently, first in the correspondingly calculated quark-photon vertex, and then consequently in the time-like electromagnetic pion form factor. The obtained results are in agreement with the available experimental data and provide further evidence for the efficacy of a vector-meson dominance model. Effects of going beyond the here employed isospin-symmetric limit are discussed. Last but not least, an outlook is provided on how to include in Poincaré-covariant Faddeev approaches the effects of intermediate resonances in the calculation of baryon properties.

1 Introduction

Quantum Chromodynamics (QCD) provides the basis for our current understanding of hadron physics. Perturbative as well as non-perturbative QCD calculations have become impressively successful in the last decades. From a theoretical point of view, QCD is a model Quantum Field Theory. It realises locality and unitarity, and being asymptotically free QCD is an ultraviolet complete Quantum Field Theory. The latter property quite obviously underlies the success of perturbative calculations for hard hadronic and semi-hadronic processes.

The perturbative description of elementary particles is essentially based on the field-particle duality which means that each field in a Quantum Field Theory is associated with a physical particle. Quite obviously, this is not the case for the elementary fields of QCD, the quarks and gluons. First of all, QCD being a gauge theory, together with the fact that only gauge-invariant quantities can be observables, rules already out that quarks and gluons can be

*e-mail: reinhard.alkofer@uni-graz.at
**e-mail: angel-aml@hotmail.com
***e-mail: helios.sanchis-alepuz@silicon-austria.com

directly observed. In a gauge theory, the rôle of fields is to implement the principle of locality. The types of different elementary fields needed in the theory is related to the structure of the gauge group, it is not, at least not directly, related to the empirical spectrum of particles. Second, what is usually referred to as "confinement" is obviously going beyond the mere issue of "only-gauge-invariant-observables", respectively, colour confinement (see, *e.g.*, refs. [1, 2]), and an especially puzzling aspect is the following: As only hadrons are produced from processes involving hadronic initial states, one has to explain that the only thresholds in hadronic amplitudes are due to the productions of other hadronic states. Consequently, singularities occurring in the amplitudes of hadronic and thus composite states are exclusively due to their hadronic substructure, or phrased otherwise, due to virtual fluctuations to hadronic resonances.

This implies not only that hadron spectroscopy and hadron structure are strongly interrelated, but, in particular, that a microscopic understanding of the effect of resonances on form factors, structure functions, etc., will shed light on the precise working of quark-hadron duality. Compelling evidence for the consequences of this duality on the qualitative as well as the semi-quantitative level has been found (for a review see, *e.g.*, Ref. [3]) and contributed to a verification of this duality which is, beyond the trivial fact of the absence of coloured states, the clearest experimental signature for confinement. The attributed perfect orthogonality of the quark and gluon degrees of freedom on the one hand and hadronic states on the other hand, and thus the perfect absence of "double-counting" when doing calculations either with quarks and gluons within QCD or within purely hadronic descriptions, is nothing else but another way to express central aspects of confinement. Basing a (confessedly impossible) exact calculation for a hadronic or semi-hadronic scattering process either on elementary QCD correlation functions or only on hadronic correlations functions needs to lead to the same result by the virtue of colour confinement, the latter being a consequence of gauge invariance. The way by which the hadronic amplitudes become effectively independent of the singularities of the elementary correlation functions is then an imprint of the dynamical realisation of confinement.

The formation of an excited hadronic bound state, being an unstable resonance, is closely related to the open decay channels. This is most evident from the analysis of time-like form factors in kinematic regions close to a resonance. To capture this interplay correctly in a QCD-based calculation of a hadron form factor, the in- (respectively, out-) going hadron and the hadronic resonance which is apparent in the form factor have to be described both in a mutual consistent manner as composite objects of quarks and gluons. Thus, a calculation of a time-like form factor from QCD, or from a microscopic model based on QCD degrees of freedom, has to overcome the challenge to treat all necessary elements on the same basis and to a sufficient degree of sophistication. Only then the result will allow conclusions on the hadron structure encoded in such a form factor.

It should be emphasised that the presented study aims for comprehending the interplay of the different features of the form factor and as they arise from the QCD degrees of freedom. It is also designed such that it can be generalised to an investigation of baryon form factors, and hereby especially the nucleons' time-like form factor. (NB: Different space-like baryon form factors have been calculated in this approach, see, *e.g.*, Refs. [5–10] for some recent respective work.) A thorough understanding of the proton time-like form factor at very low Q^2 is a very timely subject as the upcoming PANDA experiment possesses the unique possibility to measure the proton's electromagnetic form factors in the so-called unphysical region through the process $\bar{p}p \to l^+l^-\pi^0$, $l = e, \mu$ [11]. At large Q^2 the question of the onset of the convergence scale between the space-like and the time-like form factors arises.

In this talk an exploratory study of the time-like pion electromagnetic form factor employing a combination of Bethe-Salpeter and Dyson-Schwinger equations [4] will be

discussed, and its results will be presented. To obtain the most important features of the pions' time-like form factor one needs to take into account
(i) the pion as bound state of quark and antiquark thereby at the same accommodating its special role as would-be Goldstone boson of the dynamically broken chiral symmetry of QCD,
(ii) the mixing of the ρ-meson, being determined consistently as a quark-antiquark bound state, with a virtual photon when this photon is in turn coupled to a quark-antiquark pair via the fully renormalised quark-photon vertex, and
(iii) the dominant decay channel of the ρ-meson, namely, $\rho \to \pi\pi$.

Before presenting how to combine these in our study a few general remarks about the time-like pion electromagnetic form factor are in order.

2 The time-like electromagnetic pion form factor: General aspects

Figure 1. Electron-positron pair annihilating to a virtual photon with time-like momentum which then decays to a pion pair.

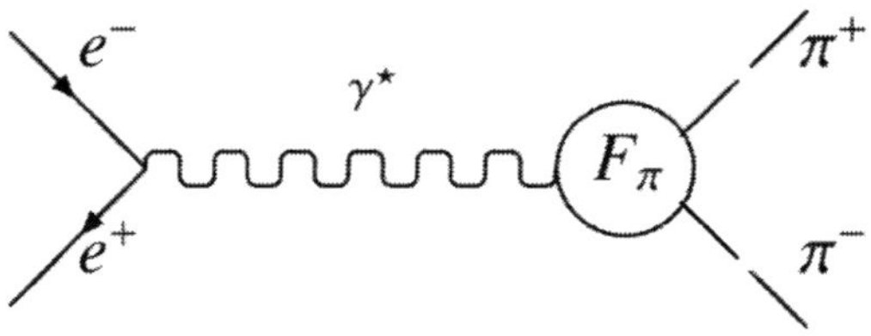

Experimentally the time-like electromagnetic pion form factor is typically determined from electron-positron pair annihilation into a pion pair, see Fig. 1. In the physical region above the two-pion production threshold the corresponding amplitude possesses a cut, and thus the time-like pion form factor has to be treated as a complex quantity.

The pion form factor contains all kind of processes turning a virtual photon into a pion pair, *cf.* Fig. 1. Using that in the interesting kinematic regime strong-interaction processes dominate by orders of magnitude against electroweak processes, an important first ingredient to the pion form factor is how the virtual photon couples to a quark via all possible QCD processes. The corresponding amplitude is nothing else than the full quark-photon vertex which includes, at least in principle, the information about the virtual photon's hadronic substructure. The two more further ingredients are the quark propagator and the pion bound state amplitude.

The presented study is performed in the isospin symmetric limit.[1] As isospin breaking is clearly visible in the time-like pion form factor, in particular through ρ-ω mixing, a vector meson dominance (VMD) based fit is used to extract the expected form of the pion form factor without this mixing effect, see Fig. 2.

As can be seen from Fig. 2 the VMD model of ref. [13] describe the experimental data precisely. But how does it arise from the dynamics of QCD degrees of freedom?

[1] Isospin breaking and thus the effect of ρ-ω mixing on the pion form factor in the presented approach is currently under investigation. Corresponding first results on electromagnetic and strong isospin breaking effects on the mass splittings of light pseudo-scalar and vector mesons calculated from Bethe-Salpeter and Dyson-Schwinger equations became recently available [15].

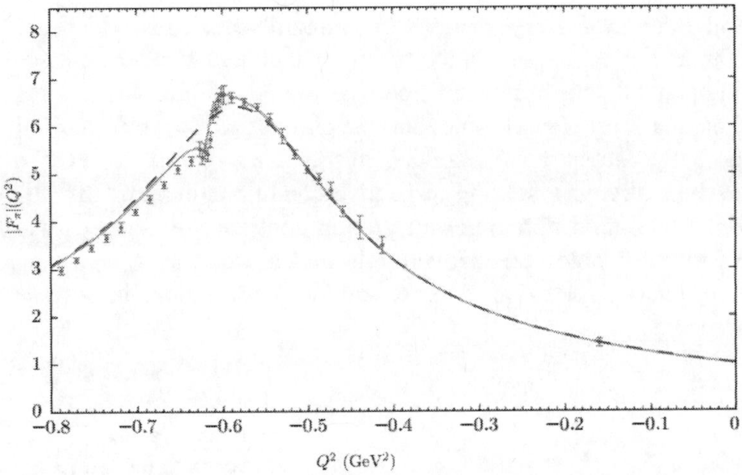

Figure 2. Absolute value of the pion form factor in the time-like ($Q^2 < 0$) domain from the VMD-based fit given in ref. [13] (full line) in comparison to the experimental data [14]. The dashed line is based on the same expression but without ρ-ω mixing, *i.e.*, the mixing matrix element is put to zero.

3 Interactions in Dyson-Schwinger and Bethe-Salpeter equations

The interactions used in the exploratory study [4] have been a gluon exchange (in the form of the Maris-Tandy model [16, 17]) as well as a pion exchange and s- and u-channel pion decay contributions, see Fig. 3, the latter both in self-consistent manner. Here the following disclaimer is in order: To keep this calculation feasible a number of technically motivated approximations have been made. It has been checked that these had only a minor impact on the results, for further details see Ref. [4].

Figure 3. Truncations used herein for the Bethe-Salpeter interaction kernel K (upper diagram) and the quark Dyson-Schwinger equation (lower diagram). In the upper diagram, the terms on the right-hand side correspond to the gluon, pion exchange, and s- and u-channel pion decay contributions to the truncation, respectively. The s- and u-channel pion decay terms do not contribute to the quark Dyson-Schwinger equation.

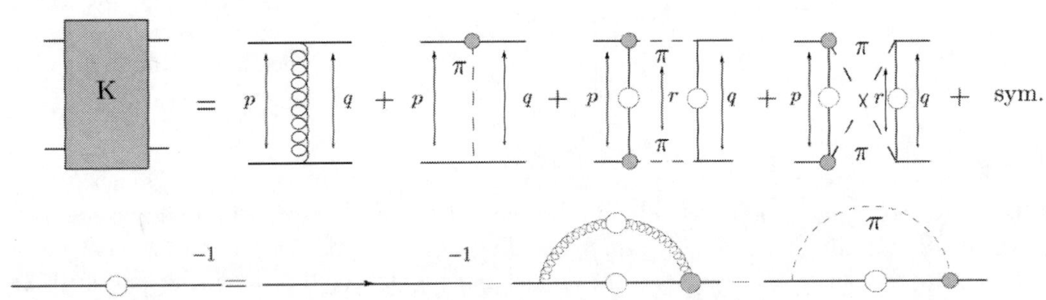

As already stated the quark-photon vertex is a decisive element in the presented calculation. It has been determined with the above mentioned interactions in Ref. [18]. The Dirac structure of the vertex can be expanded in a basis consisting of twelve elements, and all of them are taken into account to obtain the below displayed results. In addition, out of these twelve structure eight are transverse to the photon's momentum. These transverse amplitudes display a pole at the (calculated) ρ-meson mass and multi-pion cut above the two-pion threshold.[2]

[2]This quark-photon vertex has been recently used in a calculation of the pion and the kaon box contribution to a_μ^{HLbL} [19] demonstrating its usability also in the context of hadronic contributions to light-light scattering.

The Dyson-Schwinger and Bethe-Salpeter equations are integral equations which are solved numerically. Hereby, for the presented calculation the major technical challenge arises from the singularities of the integrands. This implies that one has to find integration contours in presence of cuts generated by quark propagator poles and the pion propagator pole as well as two-pion cuts and the ρ pole, the latter two also arising in the quark-photon vertex. To this end one has to determine, typically by resolving respective conditions numerically, the location of poles and cuts of the integrand. Then the integration contour is deformed such no cuts are crossed and no residues of poles are inadvertently picked up. Fig. 4 shows an example of such a contour deformation. Note that different contour deformations have to be done for different external momenta.

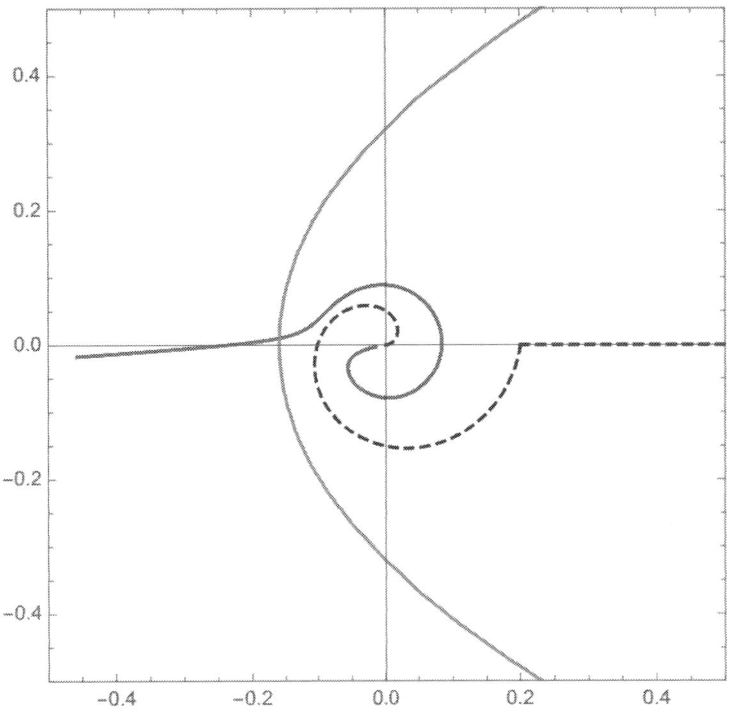

Figure 4. Shown is the complex p^2 plane such that time-like momenta correspond to the left real half-axis. Without singularities in the integrand the integration would be performed over space-like momenta, *i.e.*, over the right real half-axis. In the presented example this is not possible due to the pion cut represented by the red fill line. Correspondingly, the integration contour is deformed to the dashed blue line. The singularities of the quark propagator are all to the left of the displayed parabola, and therefore do not necessitate further contour deformations.

4 Results

For the presented exploratory calculation, the parameters are adjusted in a simple manner, especially as we aim at a qualitative understanding of the physical mechanisms involved in determining the shape of the pion form factor, and not so much at achieving an accurate quantitative agreement with experiment. As the pion-decay kernels will not only move the ρ-meson pole into the complex plane but also shift down its real value a precise fit of the two (effective) parameters of the Maris-Tandy model to reproduce the meson masses would have required quite some computational resources. In addition, only a reasonably accurate value for the pion decay constant, namely f_π= 138 MeV has resulted. Although this value is a few percent larger than the experimental one, $f_\pi^{expt.}$ = 130.5 MeV, we have adopted it in view of a compromise in the effort needed for the parameter determination and the related accuracy. In table 1 the determined meson masses as well as the pion decay constant are shown.

The pion exchange kernels lift the otherwise present degeneracy in between the iso-vector ρ- and the iso-singlet ω-meson. These interactions induce a mass splitting $m_\omega - m_\rho$ = 10 MeV which compares favourably with the experimental splitting of 7 - 8 MeV.

Table 1. The pion mass m_π, the pion decay constant f_π, the ρ-meson and ω-meson masses m_ρ and m_ω for the two different used parameterisations of the model (labeled with η) and for the case with gluon- and pion-exchange kernels but without decay kernels are shown. The current quark mass has been set to $m_q = 6.8$ MeV. The two rightmost, separated, columns display for the case when including the decay kernels the extracted ρ-meson pole position defined as $M^2_{pole} = M^2_\rho - iM_\rho\Gamma_\rho$. All values for dimensionful quantities are given in GeV.

$\Lambda = 0.78$	m_π	f_π	m_ρ	m_ω	M_ρ	Γ_ρ
$\eta = 1.5$	0.139	0.138	0.768	0.778	0.750	0.100
$\eta = 1.6$	0.126	0.138	0.774	0.784	0.759	0.105

Fig. 5 displays our results for the space-like form factor from which we also extract the result for the pion radius, $\sqrt{\langle r_\pi^2\rangle} = 0.68$ fm. It is reassuring that the space-like form factor, including the pion radius, shows a very good agreement with experimental data. As we included the most important physical processes with effects in the time-like regime we take this as evidence that the imaginary part in the time-like region is precisely enough reproduced to provide very good results for the space-like form factor.

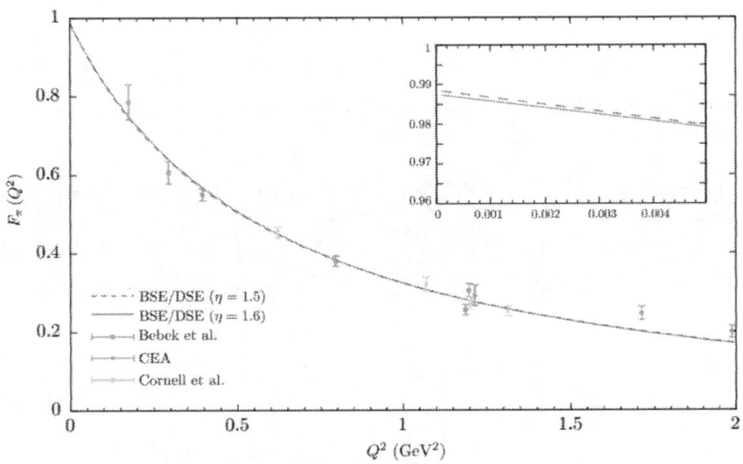

Figure 5. Pion form factor in the spacelike $Q^2 > 0$ domain for the model parameters $\eta = 1.5$ and $\eta = 1.6$. (The inset illustrates the impact of one of the technically motivated approximations.)

The results for the absolute value of the pion form factor for time-like ($Q^2 < 0$) virtualities are displayed in Figs. 6. As a consequence of the decay kernels, the pion form factor (as well as the quark-photon vertex) possess a cut along the real negative axis starting at the two-pion threshold $Q^2 = -4m_\pi^2$, and therefore also an imaginary part contributes significantly to the absolute value.

With decay kernels included, the ρ-meson pole, present in the quark-photon vertex, moves into the complex plane, and consequently the pion form factor develops a bump-like shape with approximately correct height and width. Therefore the presented calculation corrects a deficiency of earlier respective studies of the time-like pion form factor, without or even with the pion exchange term, for which the form factor diverges at the Q^2-value corresponding to the ρ-meson mass.

In order to provide a measure for the proximity of our results to the VMD predictions we performed rational fits (Padé fits) to the results for the real part and for the imaginary part of the form factor and compare the coefficients to the VMD expression without ρ-ω mixing. To

Figure 6. Absolute value of the pion form factor in the time-like $Q^2 < 0$ domain for the model parameters $\eta = 1.5$ and $\eta = 1.6$ as described in the text and compared to experimental data [14].

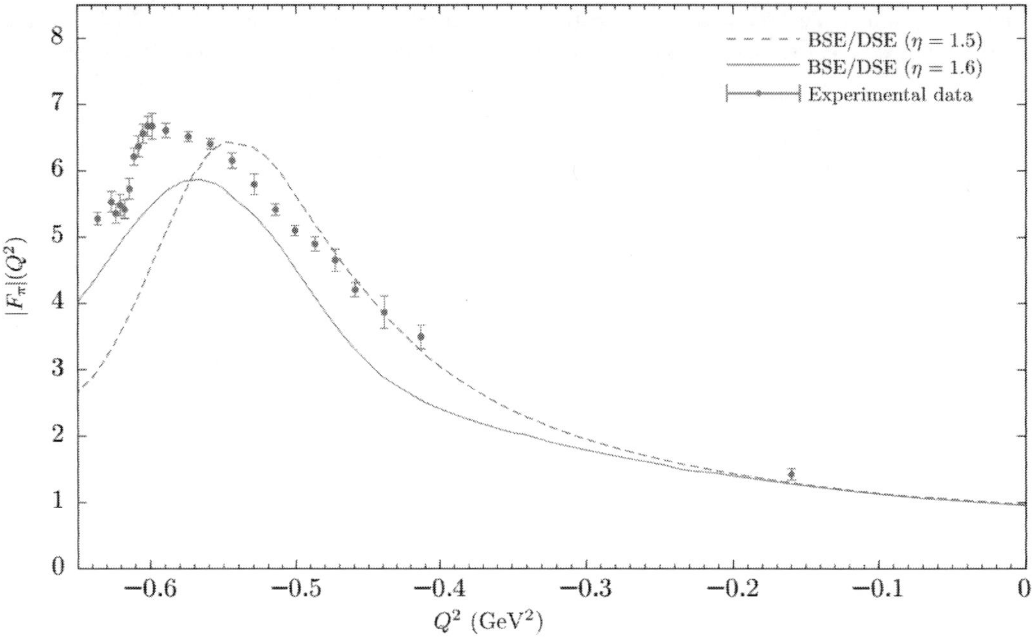

Table 2. The coefficients of the rational fits to the pion form factor as discussed in the text. All values for dimensionful quantities are given in GeV.

	η =1.5	η =1.6	VMD expression		η =1.5	η =1.6	VMD expression
a_1	0.5587	0.4149	0.72	c_1	0.0591	0.0997	0
a_2	0.8828	0.6827	1.2	c_2	0.1295	0.2383	0.2308
b_0	0.3600	0.3600	0.36	d_0	0.3600	0.3600	0.36
b_1	1.2307	1.2517	1.2	d_1	1.1924	1.2464	1.2
b_2	1.0722	1.1000	1.037	d_2	0.9973	1.0916	1.037

stay as general as possible we used first an ansatz with twelve parameters in total:

$$\begin{aligned} Re\, F_\pi(Q^2) - F_\pi(0) &\approx -\frac{a_0 + a_1 Q^2 + a_2 (Q^2)^2 + a_3 (Q^2)^3}{b_0 + b_1 Q^2 + b_2 (Q^2)^2 + b_3 (Q^2)^3} \\ Im\, F_\pi(Q^2) &\approx \frac{c_0 + c_1 Q^2 + c_2 (Q^2)^2 + c_3 (Q^2)^3}{d_0 + d_1 Q^2 + d_2 (Q^2)^2 + d_3 (Q^2)^3} . \end{aligned} \tag{1}$$

We obtained almost vanishing values for the coefficients a_0, a_3, b_3, c_0, c_3 and d_3. Phrased otherwise, *the fit confirms the structure expected from the VMD expression.* Based on this, the fits have been repeated for

$$\begin{aligned} Re\, F_\pi(Q^2) - F_\pi(0) &\approx -\frac{a_1 Q^2 + a_2 (Q^2)^2}{b_0 + b_1 Q^2 + b_2 (Q^2)^2} , \\ Im\, F_\pi(Q^2) &\approx \frac{c_1 Q^2 + c_2 (Q^2)^2}{d_0 + d_1 Q^2 + d_2 (Q^2)^2} . \end{aligned} \tag{2}$$

The coefficients resulting from these fitting ansätze as well as the ones resulting from VMD expression are given in table 2. Note that the expression based on the VMD hypothesis is an astonishingly good representation of the results based on the microscopic model of QCD degrees of freedom: Other terms than VMD-predicted ones are tiny, and the elaborated calculation yields within error margin the VMD predicted functional form.

Another remark applies to the fact that there is no significant impact from quark propagator poles on the pion form factor. As discussed in the Introduction this is a wanted feature in the view of confinement, however, it should be understood why this happens in the presented study which contains, despite its quite high level of sophistication, not all QCD features.

Reassuring is nevertheless that the resulting time-like pion form factor in the kinematic region $0 > Q^2 > 0.8\mathrm{GeV}^2$ is determined by the ρ-meson pole and the two-pion cut despite basing the calculation on QCD degrees of freedom.

5 Conclusions and Outlook

The main message of the presented study is the feasibility of studying time-like quantities in a Dyson-Schwinger–Bethe-Salpeter approach including intermediate resonances and thus decay channels in the interaction kernels. Despite the applied modelling (which can be overcome in future investigations) and the technical limitations inherent to the approach we found a remarkable agreement with the experimental data.

As a side result we provided a detailed quantitative verification of the VMD model for the pion form factor. In view of the fact that the ρ-meson pole and the two-pion cut are determining the prominent features of the pion form factor it might be considered expected that the VMD hypothesis should be, at least, qualitatively correct. Therefore, the more surprising, and definitely positive, aspect is that a microscopic model built on QCD degrees of freedom provide a pion form factor shaped by the hadron substructure of the virtual photon.

One obvious way to improve the presented study is to take isospin breaking via the different quark masses and electric charges into account. First steps in this direction have been undertaken [15], however, a microscopic description for ρ-ω-mixing including the decay channels for these vector mesons, and its implementation into a calculation of the pion form factor, is still a task to be performed yet.

It would be also worth to extend the presented study to calculate the $\gamma\,\pi\,\pi\,\pi$ form factor which is only non-vanishing due to the chiral anomaly. How in such a form factor, whose soft-point value is fixed due to the anomaly and thus easily understood from the perspective of QCD, the hadron resonances are determining the prominent time-like features will be of general interest.

Last but not least, the question of the accessibility of the nucleons' time-like form factors arises. The presented study paves the way for an analogous inclusion of decay kernels into baryon bound state equations. To this end we note that a functional approach based on Dyson-Schwinger and bound-state equations allows for a unified description of mesons and baryons, see, *e.g.*, Ref. [5], which quite successfully describes the hadron spectrum and space-like form factors. However, exactly these studies provide an estimate about the dramatic increase in complexity when dealing with baryon instead of meson bound state equations and a respective calculation of form factors based on these equations. Nevertheless, given the increased effort from the experimental side and the resulting anticipated highly precise data over a large kinematical region certainly justifies the correspondingly increased effort on the theoretical side.

Acknowledgements

We are grateful to the organisers of the *International Conference on Exotic Atoms and Related Topics* (EXA21) for all their efforts which made eventually this (online) conference possible. This work was partially supported by the the Austrian Science Fund (FWF) under project number P29216-N36.
A.S. Miramontes acknowledges CONACyT for financial support.
The numerical computations have been performed at the high-performance compute cluster of the University of Graz.

References

[1] J. Greensite, *An introduction to the confinement problem*, Lect. Notes Phys. **972**, 2nd ed., 2020.
[2] R. Alkofer and J. Greensite, J. Phys. G **34** (2007) S3 [arXiv:hep-ph/0610365 [hep-ph]].
[3] W. Melnitchouk, R. Ent and C. Keppel, Phys. Rept. **406** (2005) 127 [arXiv:hep-ph/0501217 [hep-ph]].
[4] Á. Miramontes, H. Sanchis-Alepuz and R. Alkofer, Phys. Rev. D **103** (2021) 116006 [arXiv:2102.12541 [hep-ph]].
[5] G. Eichmann, H. Sanchis-Alepuz, R. Williams, R. Alkofer and C. S. Fischer, Prog. Part. Nucl. Phys. **91** (2016) 1 [arXiv:1606.09602 [hep-ph]].
[6] D. Nicmorus, G. Eichmann and R. Alkofer, Phys. Rev. D **82** (2010) 114017 [arXiv: 1008.3184 [hep-ph]].
[7] G. Eichmann, Phys. Rev. D **84** (2011) 014014 [arXiv:1104.4505 [hep-ph]].
[8] G. Eichmann and C. S. Fischer, Eur. Phys. J. A **48** (2012) 9 [arXiv:1111.2614 [hep-ph]].
[9] H. Sanchis-Alepuz, R. Williams and R. Alkofer, Phys. Rev. D **87** (2013) 096015 [arXiv:1302.6048 [hep-ph]].
[10] H. Sanchis-Alepuz, R. Alkofer and C. S. Fischer, Eur. Phys. J. A **54** (2018) 41 [arXiv:1707.08463 [hep-ph]].
[11] G. Barucca *et al.* [PANDA], [arXiv:2101.11877 [hep-ex]].
[12] B. Ananthanarayan, I. Caprini and D. Das, Phys. Rev. D **102** (2020) 096003 [arXiv: 2008.00669 [hep-ph]].
[13] H. B. O'Connell, B. C. Pearce, A. W. Thomas and A. G. Williams, Prog. Part. Nucl. Phys. **39** (1997) 201 [arXiv:hep-ph/9501251 [hep-ph]].
[14] R. R. Akhmetshin *et al.* [CMD-2], Phys. Lett. B **648** (2007) 28 [arXiv:hep-ex/0610021 [hep-ex]].
[15] Á. Miramontes, R. Alkofer, C. S. Fischer and H. Sanchis-Alepuz, in preparation.
[16] P. Maris and C. D. Roberts, Phys. Rev. C **56** (1997) 3369 [nucl-th/9708029].
[17] P. Maris and P. C. Tandy, Phys. Rev. C **60** (1999) 055214 [nucl-th/9905056].
[18] Á. S. Miramontes and H. Sanchis-Alepuz, Eur. Phys. J. A **55** (2019) 170 [arXiv: 1906.06227 [hep-ph]].
[19] Á. Miramontes, A. Bashir, K. Raya and P. Roig, arXiv:2112.13916 [hep-ph].

Feasibility study of tau-lepton anomalous magnetic moment measurements with ultra-peripheral collisions at the LHC

Paul Bühler[1,*], *Nazar* Burmasov[2], *Roman* Lavička[1], and *Evgeny* Kryshen[2]

[1]Stefan Meyer Institute for Subatomic Physics, Austrian Academy of Sciences, Vienna, Austria
[2]Petersburg Nuclear Physics Institute named by B. P. Konstantinov of National Research Center «Kurchatov Institute», Gatchina, Russia

Abstract.
Precision measurements of the anomalous electromagnetic moment of leptons (a_l) may serve as one of the most promising directions in the search for new physics beyond the Standard Model. While the experimental value of the electron magnetic moment agrees with theoretical predictions with up to 11 significant digits, the muon magnetic moment shows deviations from the Standard Model value at the level of 4.2 sigma, indicating the possible occurrence of new physics effects. Although the a_τ of the tau lepton with its heavy mass is expected to be $m_\tau^2/m_\mu^2 \approx 280$ times more sensitive to new physics effects than a_μ, measurements of this quantity are rare. This is because the standard spin precession methods are not suitable for a_τ measurements due to the very short tau lifetime.
Ultra-peripheral collisions of heavy ions at the LHC may serve as an alternative tool to measure a_τ. In ultra-peripheral collisions, hadronic interactions are strongly suppressed and long-distance electromagnetic processes dominate, providing an environment to study the electromagnetic properties of the tau lepton. The di-tau production process PbPb $\rightarrow$ PbPb$\gamma\gamma$ $\rightarrow$ PbPb$\tau\tau$ contains two gamma-tau vertices and hence provides enhanced sensitivity to the anomalous magnetic and electric moments.
In this contribution we discuss the feasibility of the a_τ measurement in ultra-peripheral collisions with the ALICE experiment and present projections of the sensitivity of the measurement for the upcoming heavy ion run in 2022 at LHC.

1 Introduction

The magnetic moment $\vec{\mu}_S$ of a charged elementary particle is related to its spin $\vec{S}$ by $\vec{\mu}_S = g \cdot \mu_B \cdot \vec{S}$, where g is the dimensionless g-factor and μ_B the Bohr magneton. The measurement of the lepton magnetic moments μ_l, with $l = e, \mu, \tau$ provides a sensitive test of the Standard Model (SM). A precision measurement is of interest because their SM values can be calculated with high precision and any deviation of the experimental value from the predicted SM values would be an indication for effects of Beyond the Standard Model (BSM) physics. Traditionally, the comparison is made in terms of the anomalous magnetic moment

*e-mail: paul.buehler@oeaw.ac.at

a_l, which is defined as $a_l = \frac{g_l-2}{2}$. Since g_l is approximately 2, a_l is expected to have a value close to 0.

The SM values of the lepton anomalous magnetic moments can be calculated with high precision. In general, they contain contributions from all domains of the SM, from QED, the weak, and the strong interaction. State-of-the-art calculations are discussed in Refs. [1] (electron), [2] (muon), and [3] (tau). Whereas for the electron the "anomaly" is determined by QED effects, contributions by the weak and strong forces are important in the case of the muon and tau. In addition so far unknown BSM particles could contribute as well. Not only in the SM, but also in many new physics models, additional particles with mass M contribute to the theory with terms typically proportional to m_l^2/M^2. For the three charged leptons this implies a relative sensitivity ϵ_l to BSM effects of $\epsilon_e : \epsilon_\mu : \epsilon_\tau = 1 : 42'754 : 12'092'970$, which is a strong motivation for a detailed study of a_τ.

One common method to measure the magnetic moment of a charged particle is by observing its precession in a magnetic field, of which the frequency is directly proportional to the magnetic moment. In order to perform the measurements the leptons must be stored for a long enough time to be able to measure its precession frequency. Such experiments have been realized for electrons [4] and muons [5] with remarkable precision. Within the achieved precision of 11 significant digits the value determined for a_e is in agreement with the SM value and in case of the a_μ it is still to be clarified weather the difference of 4.2 standard deviations between the currently most precise measurement and theory value is an indication of BSM physics or weather this tension is solved by refined calculations [6] or improved measurements [7].

Due to the short lifetime of the tau lepton, the precession method is not suitable to determine its anomalous magnetic moment. For this, however, the fact that the coupling strength of a tau to a pair of photons, $\Gamma_\mu^{\gamma ll}$ depends on a_τ can be exploited. Hence cross sections of reactions including such a vertex are sensitive to the value of a_τ and their precise determination allows to test for deviations of a_τ from its SM value. The photon-lepton vertex function can be parameterized as [8]

$$\Gamma_\mu^{\gamma ll}(q) \propto \gamma_\mu F_1(q^2) + \frac{1}{2m_\tau}(iF_2(q^2) + \gamma_5 F_3(q^2))\sigma_{\mu\nu}q^\nu, \tag{1}$$

where q is the momentum transfer $p' - p$, where p and p' are the four-momenta of the incoming and outgoing leptons, respectively. $\sigma_{\mu\nu} = \frac{i}{2}[\gamma_\mu, \gamma_\nu]$ is the spin $1/2$ angular momentum tensor, and the γ_k are the Dirac-Matrices. In the static limit where $q = 0$ the form factors $F_{1,2,3}$ are related to the electromagnetic properties of the lepton by $q_l = F_1(q^2 = 0)$, $a_l = F_2(q^2 = 0)$, and $d_l = 2m_l/eF_3(q^2 = 0)$, where q_l is the charge in units of e and d_l the electric dipole moment of the lepton. This results in a quite sizable a_τ dependence of the elementary $\gamma\gamma \to \tau^+\tau^-$ cross sections as shown in Figure 1.

The best measurement of a_τ has been made by the DELPHI collaboration by measuring the cross section of the reaction $e^+e^- \to e^+e^-\gamma\gamma \to e^+e^-\tau^+\tau^-$ at LEP2 [10]. The experimental and theoretical values of the lepton anomalous magnetic moments are summarized in Table 1. Whereas in the case of the electron and muon the precision of the experimental and theoretical value are of the same order of magnitude, this is different in the case of the tau lepton, where the measured value lacks precision.

A number of proposals for improvement of the experimental limits have been put forward including measurements at Belle-II [11], LHC [12] and future facilities like CLIC [13], LHeC and FCC-he [14] or high-luminosity B factories [15]. The possibility to determine a_τ in heavy-ion ultra-peripheral collisions (UPCs) was pointed out by del Aguila et al. [16] already in 1991. Recently, this idea has been taken up and feasibility studies to perform this measurement at LHC with the CMS and ATLAS experiments have been presented [17, 18].

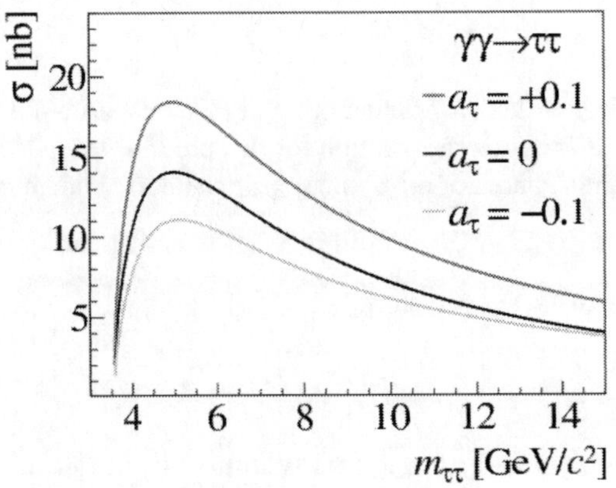

Figure 1. Cross section of the $\gamma\gamma \to \tau^+\tau^-$ reaction as function of the 2τ invariant mass $m_{\tau\tau}$ for different values of the tau anomalous magnetic moment a_τ [9].

Table 1. Theoretical and experimental values of the lepton anomalous magnetic moments.

lepton	theory		experiment	
electron	$1159652181.606(11) \times 10^{-12}$	[1]	$1159652180.91(26) \times 10^{-12}$	[20]
muon	$116591810(43) \times 10^{-11}$	[2]	$116592061(41) \times 10^{-11}$	[5]
tau	$117721(5) \times 10^{-8}$	[3]	$-0.052 < a_\tau < 0.013$ (95%C.L.)	[10]

In the remaining of this paper we will discuss the method and the possibility to measure a_τ with the ALICE experiment in UPCs.

2 Measurement of a_τ with UPCs

Ultra Peripheral Collisions are collisions of heavy ions in which the impact parameter b is larger than the sum of the radii of the colliding projectiles. Due to the large impact parameter hadronic reactions are heavily suppressed and electromagnetic interactions dominate. In UPCs, the ions flying past each other, produce strong electromagnetic fields that effectively form intense fluxes of energetic photons. The flux of photons $n(w)$ of a given energy w can be calculated using the equivalent photon approximation (EPA) [19]. The photons are almost real ($q^2 \sim 0$) and the spectrum is rather soft, making UPCs an ideal place to study the electromagnetic properties of the tau lepton.

Here we consider the photoproduction of di-tau in the reaction PbPb $\to$ PbPb$\tau^+\tau^-$ which is depicted in Figure 2. Since the photon flux from an ion scales with Z^2, where Z is the charge of the ion, the related cross section is expected to scale with Z^4. The UPC cross section, σ_{PbPb}, is given by the convolution of the photon spectra with the elementary photon-photon cross section $\sigma_{\gamma\gamma}$ by[1]

$$\sigma_{PbPb} = \int_0^\infty \int_0^\infty \sigma_{\gamma\gamma} \cdot n_{Pb}(w_1) \cdot n_{Pb}(w_2)\, dw_1 dw_2. \quad (2)$$

In order to measure the cross section of the above reaction one needs to detect the decay products of the outgoing taus. The dominating decay channels of the tau are listed in Table 2 together with the respective branching ratios. All decay channels include neutrinos which are not detected with the apparatus at LHC. The analysis strategy is thus not the exclusive

[1] A more precise approach takes into account the impact parameter dependence of the photon fluxes and the probabilitiy of hadronic interactions between colliding nuclei [9].

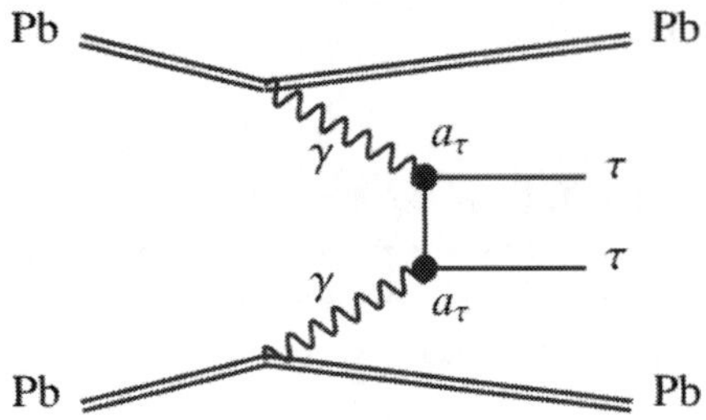

Figure 2. Pair production of tau leptons in Pb–Pb UPCs. The cross section for this process depends on the value of the anomalous magnetic moment a_τ.

reconstruction of the events but the detection of events with a lepton (electron or muon) plus one or three additional charged particles and otherwise empty detectors. Possible background reactions from hadronic interactions are effectively suppressed by the few-prong condition. Exclusive production of di-electron/muon pairs, an other potential source of background, is distinguished by the back-to-back emission of the two leptons. Semi-coherent di-lepton production in which the momentum transfer from a gamma to an ion leads to the dissociation of the ion, is accompanied by emission of forward neutrons, which are detectable in the Zero Degree Calorimeters (ZDC) of the systems.

Table 2. Dominating decay channels of tau leptons [20].

Decay channel	Branching ratio [%]
$\tau^\pm \rightarrow e^\pm \nu_e \nu_\tau$	17.8
$\tau^\pm \rightarrow \mu^\pm \nu_\mu \nu_\tau$	17.4
$\tau^\pm \rightarrow \pi^\pm \nu_\tau + n\pi^0$	45.6
$\tau^\pm \rightarrow$ 3 prong	~ 20

3 ALICE potential for Run 3

ALICE has gained quite some experience with UPC physics during Run 1 & 2 and its detector is well equipped for this (see e.g. [22]). During the last three years, during the long shutdown at LHC, the ALICE experiment has been upgraded to cope with a continuous data acquisition mode for Run 3 & 4 [21] which will allow to acquire Pb–Pb collisions at a maximum rate of 50 kHz. For the analysis of UPCs, ALICE profits of many of its sub-detectors. The Inner Tracking System (ITS) and Time Projection Chamber (TPC) are the central tracking detectors. They allow to identify charged particles with transverse momenta down to a few hundred MeV in the pseudorapidity range of $|\eta| < 0.9$. The region at $-4 < \eta < -2.5$ is covered by a muon arm. With the recently installed Muon Forward Tracker a precise determination of the production vertex will be possible allowing to discriminate muons produced in secondary vertices. The Fast Interaction Trigger detector (FIT) consists of several sub-detectors which are not only important for luminosity measurements, but also allow to veto additional activity over a large η range. Whereas the electrons can be well identified in the central region of ALICE, it is more difficult to distinguish muons from pions due to their similar mass and hence similar energy deposit in the sensitive parts of the detector. In the range covered by the muon arm, muons can be identified. Hence in ALICE two types of events will be considered for this measurement - events with an identified electron (leading electron) and an additional charged track in the central barrel (central event type) and events with a muon in forward direction plus a charged track in the central barrel (semi-forward event type).

It is planed that in Fall 2022 ALICE will accumulate an integrated luminosity of 2.7 nb^{-1} of Pb–Pb collisions at $\sqrt{s_{NN}} = 5.5$ TeV. In order to study the ALICE potential to measure a_τ with this data we performed simulations with the event generator Upcgen [9]. The estimated yield of central events in the ALICE fiducial acceptance is estimated to be around 35′000 and the yield of semi-forward events, approximately 2′000. In Figure 3 the yield of central events is plotted as function of p_T of the leading electron for different values of a_τ.

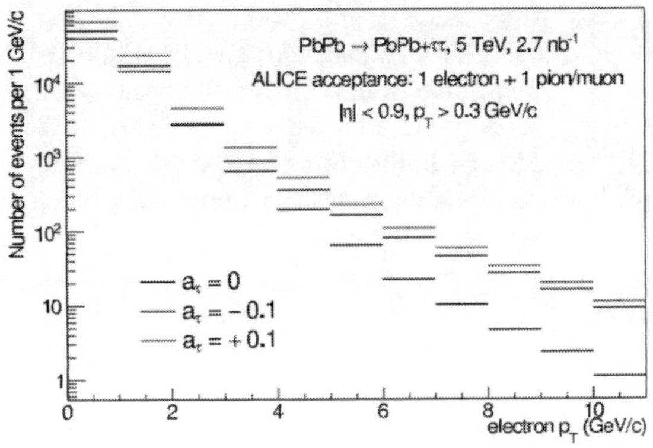

Figure 3. Number of PbPb → PbPb$\tau^+\tau^-$ events expected to be seen by the ALICE detector during the Fall 2022 Pb–Pb run at LHC. Here at least one of the taus is assumed to decay to $e^\pm\nu_e\nu_\tau$. The yield is plotted as function of the leading electron p_T for three different values of a_τ. Both, the absolute value and the slope of the yield function are sensitive to a_τ.

The p_T spectrum does not only simply scale with the value of a_τ but also changes slope. This is demonstrated with Figure 4 where the relative changes of the yields are plotted as function of a_τ for three different p_T bins of the leading electron. Whereas the a_τ dependence at low p_T is approximately linear, it becomes parabolic at higher momenta. This complex p_T-differential behavior of the a_τ dependence can be exploited to enhance the sensitivity of the measurement.

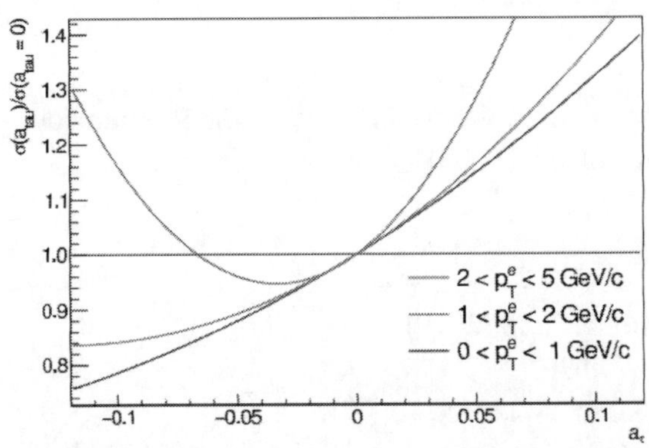

Figure 4. Relative change of central events as function of a_τ in three distinct bins of the leading electron p_T. Whereas the a_τ dependence at low p_T is approximately linear, it becomes parabolic at higher momenta.

The upper limit on a_τ, which can be set by the ALICE collaboration with the data sample from the first year of Run 3, are estimated using the following definition of χ^2

$$\chi^2(a_\tau) = \sum_i \frac{(S_i(a_\tau) - S_i(0))^2}{S_i(0) + \zeta S_i^2(0)}, \tag{3}$$

where $S_i(a_\tau)$ is the yield of leading electrons in the p_T bin i and ζ is the expected level of the relative systematic uncertainty. Background contributions are neglected here. Expected limits on a_τ at 68% and 95% CL are shown in Figure 5 for three different assumptions about the systematic uncertainty ζ (1%, 3%, 5%). The systematic errors will be dominated by the uncertainty of the luminosity determination and track reconstruction. For comparison, the DELPHI results and Standard Model Effective Field Theory (SMEFT) predictions corresponding to the range of confinement scales in a composite tauon scenario [17] are also included in the figure.

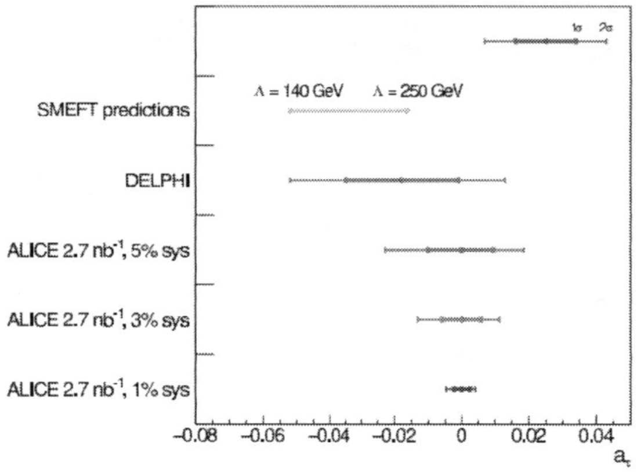

Figure 5. Comparison of the limits on a_τ from DELPHI and expected from the first year of Run 3 data taking with ALICE. The final value from ALICE will depend on the value of the achieved systematic errors. The SMEFT limits correspond to the range of confinement scales in a composite tau scenario.

4 Conclusion

UPCs at LHC provide a unique opportunity to study electromagnetic processes. In particular, with data from Run 3 limits of the anomalous magnetic moment of the tau lepton can be determined by measuring the cross section of the di-tau photoproduction in the PbPb $\to$ PbPb$\tau\tau$ process. Depending on the systematic errors which will be achieved, the data accumulated in the first year of Run 3 will allow ALICE to set new limits which are significantly below the existing best value published by the DELPHI collaboration in 2004.

Acknowledgments
This work is supported by the Austrian Science Fund (FWF, I 5277-N) and the Russian Foundation for Basic Research (RFBR, project No. 21-52-14006).

References

[1] T. Aoyama, T. Kinoshita, and M. Nio, Atoms **7**, 28 (2019).
[2] T. Aoyama et al., Physics Reports **887**, 1 (2020).
[3] S. Eidelman and M. Passera, Mod. Phys. Lett. A **22**, 159 (2007).
[4] D. Hanneke, S. Fogwell Hoogerheide, and G. Gabrielse, Phys. Rev. A **83**, 052122 (2011).
[5] B. Abi et al., Phys. Rev. Lett. **126**, 141801 (2021).
[6] Sz. Borsanyi et al., Nature **593** (2021) 51.
[7] M. Abe et al., Prog. Theor. Exp. Phys. **2019** (2019) 053C02.
[8] M. E. Peskin and D. V. Schroeder, CRC Press, ISBN-13: 978-0-201-50397-5 (1995).
[9] N. Burmasov et al., arXiv:2111.11383 (2021).
[10] The DELPHI collaboration, Eur. Phys. J. C **35**, 159 (2004).
[11] X. Chen and Y. Wu, J. High.Ener.Phys. **2019**, 10 (2019).
[12] A. S. Fomin et al., J. High.Ener.Phys. **2019**, 3 (2019).
[13] A.A. Billur et al., Phys. Rev. D **98**, 095013 (2018).

[14] M. Köksal, J. Phys. G **46**, 065003 (2019).
[15] A. Crivellin, M. Hoferichter, and J. M. Roney, arXiv:2111.10378 [hep-ph] (2021).
[16] F. del Aguila, F. Cornet, and J.I. Illana, Phys. Lett. B **271**, 256 (1991).
[17] L. Beresford and J. Liu, Phys. Rev. D **102**, 11 (2020).
[18] M. Dyndał et al., Phys. Lett. B **809**, 135682 (2020).
[19] C. F. V. Weizsäcker, Z.Physik **88**, 612 (1934); E. J. Williams, Kgl. Danske Vidensk. Selskab. Mat.-Fiz. Medd. **13**, 4 (1935).
[20] P.A. Zyla et al. (Particle Data Group), Prog. Theor. Exp. Phys. **2020**, 083C01 (2020) and (2021) update.
[21] ALICE collaboration, J. Phys. G **41** 087001 (2014).
[22] ALICE collaboration, Eur. Phys. J. C **79** 5 (2019), Phys. Lett. B **798** 134926 (2019), JHEP **6** 35 (2020), Phys. Lett. B **817** 136280 (2021), Eur. Phys. J. C **81** 712 (2021).

Analysis methods used and planned for VIP-2

Alessio Porcelli[1,2,*], *Sergio* Bartalucci[2], *Sergio* Bertolucci[3], *Massimiliano* Bazzi[2], *Mario* Bragadireanu[2,4], *Cesidio* Capoccia[2], *Michael* Cargnelli[1], *Alberto* Clozza[2], *Catalina* Curceanu[2], *Luca* De Paolis[2], *Raffaele* Del Grande[5,2], *Carlo* Fiorini[6], *Carlo* Guaraldo[2], *Mihai* Iliescu[2], *Matthias* Laubenstein[7], *Johann* Marton[1], *Marco* Miliucci[2], *Edoardo* Milotti[8], *Fabrizio* Napolitano[2], *Kristian* Piscicchia[9,2], *Alessandro* Scordo[2], *Hexi* Shi[1], *Diana Laura* Sirghi[2,4], *Florin* Sirghi[2,4], *Francesco* Sgaramella[2], *Oton* Vazquez Doce[2,5], and *Johann* Zmeskal[1]

[1]Stefan Meyer Institute for Subatomic Physics, Kegelgasse 27, 1030 Wien, Austria
[2]INFN, Laboratori Nazionali di Frascati, Via E. Fermi 54, I-00044 Frascati(RM), Italy
[3]Dipartimento di Fisica e Astronomia, Università di Bologna and INFN—Sezione di Bologna, Via Irnerio 46, I-40126 Bologna, Italy
[4]Horia Hulubei National Institute of Physics and Nuclear Engineering, Str. Atomistilor No. 407, P.O. Box MG-6 Buchares-Magurele, Romania
[5]Excellence Cluster Universe, Technische Universität München, Boltzmannstraße 2, 85748 Garching bei München, Germany
[6]Politecnico di Milano, Dipartimento di Elettronica, Informazione e Bioingegneria and INFN Sezione di Milano, I-20133 Milano, Italy
[7]INFN, Laboratori Nazionali del Gran Sasso, Via G. Acitelli 22, I-67100 Assergi - L'Aquila, Italy
[8]Dipartimento di Fisica, Università di Trieste and INFN—Sezione di Trieste, Via Valerio, 2, I-34127 Trieste, Italy
[9]Centro Ricerche Enrico Fermi—Museo Storico della Fisica e Centro Studi e Ricerche "Enrico Fermi", Via Panisperna 89a, 00184 Roma, Italy

Abstract. VIP-2 (VIolation of Pauli exclusion principle - 2) is an underground experiment sited in the underground "Laboratori Nazionali del Gran Sasso." It aims to investigate possible violations of the Pauli Exclusion Principle (PEP) and, in this context, Quantum Gravity models implying violations of PEP. While an upper limit of PEP violation probability is recently published, the data requires further developments of accurate analysis techniques and methods. In this contribution, we present an overview of the methodologies proposed for current and planned analysis.

1 Introduction

As a fundamental pillar of the quantum mechanics for the stability of atoms, Wolfgang Pauli in 1943 formulated the Pauli Exclusion Principle (PEP). It is a direct implication of the Spin Statistic Theorem (SST) [1]. Particles in the Standard Model are grouped in two categories: the bosons with integer spin and the fermions with half-integer spin. Within Quantum Field Theory (QFT), they transform differently by exchanging identical particles: bosonic states have symmetrical permutations, and fermionic ones have antisymmetrical [2].

*e-mail: alessio.porcelli@oeaw.ac.at

In a given system of identical particles, transitions are forbidden between different symmetry states by the Messiah-Greenberg superselection rule (MG) [3]. The Pauli exclusion principle prohibits two electrons with the same quantum spin number from simultaneously occupying the same atomic energy level in the system. Consequently, more than two electrons cannot occupy each atomic energy level with the opposite quantum number of spin. In such a way, any atomic transition to an energy level already occupied by two electrons implies a PEP violation. Accordingly to MG, a PEP-violating signal has to be searched in open quantum systems, i.e., by looking for transitions among violating states of a prepared system after introducing particles from outside the considered system. Transitions among anomalous states would occur at the standard rate if the involved particles couple universally to the interaction field. An Open Systems experiment tests PEP violation ([4, 5]) constrained by the MG superselection rule, opening a door for a new physics behind the Standard Model.

VIP-2 (VIolation of Pauli exclusion principle - 2) is an Open Systems experiment testing the Pauli Exclusion Principle (PEP), injecting electrons in copper. It measures copper atomic transitions with high precision X-rays detectors and looks for a PEP-violating Kα transition (2p → 1s). The energy of the PEP-forbidden Kα atomic transition (schematized in figure 1) would be shifted by about 300 eV, with respect to the standard one, due to the extra shielding provided by the two electrons in the 1s state of the atom [6].

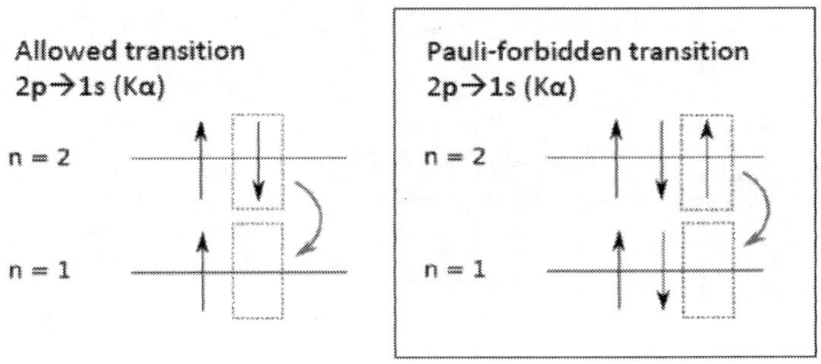

Figure 1. Schematic of PEP-allowed and PEP-violating Kα transition, respectively, on the left and the right.

The number of the PEP-violating emitted X-rays N_x is estimated as follows [7]:

$$N_x \sim \frac{\beta^2}{2} \cdot N_{\mathrm{new}} \cdot \frac{N_{\mathrm{int}}}{10} \cdot \mathrm{efficiency}\,, \tag{1}$$

where $\beta^2/2$ is the PEP-violating probability, N_{new} is the total number of "new" conduction electrons injected into the system, the factor 1/10 is an estimate of the capture probability (per scattering) into the 2p state, N_{int} is the minimum number of electron-atom interactions as electrons flow in the target, and the "efficiency" takes into account both the solid angle covered by the detector and the X-ray absorption in the copper strip.

The goal of the VIP-2 experiment is to improve the upper limit on the Pauli exclusion principle violation probability $\beta^2/2 < 4.7 \times 10^{-29}$, previously set by the VIP experiment, by two orders of magnitude. We present the VIP-2 analysis methodologies used and planned to achieve the aimed goal in this present work.

VIP-2 is also used in a Closed Systems mode. Recent calculations suggest that spin-statistics violations may emerge naturally in the Quantum Gravity models [8, 9], not subject to the MG superselection rule. Consequently, they can also be tested using this configuration without introducing test particles from outside the system. The VIP-2 collaboration has recently undertaken an effort in this direction, but it will not be treated in this work.

2 VIP-2

The VIP-2 experimental apparatus was transported and mounted in the Laboratori Nazionali del Gran Sasso (LNGS) at the end of 2015. Sited beneath the Gran Sasso mountain, the setup is shielded from cosmic rays. The first data-taking campaign started in October 2016 as part of the commissioning phase and ended in November 2017. The detector consisted of two arrays of 1×3 Silicon Drift Detectors (SDDs) surrounding the copper target, each array with 3 cm^2 of the effective surface. The experimental setup was further upgraded in April 2018, when the detector system was replaced with two new arrays, each with 2×8 SDDs, for a total of 32 SDDs. The SDDs are cooled down to about -90 °C. The target system is cooled down by water cooling, keeping them to a temperature of about 12 degrees. When the current of 100 A is turned on, the strips' temperature increases by a few degrees. That induces a temperature rise at the SDDs of the same amount, which does not significantly alter the SDDs' performances. A schematic view of the VIP-2 apparatus is shown in figure 2. Currently published results use the data taken in this configuration.

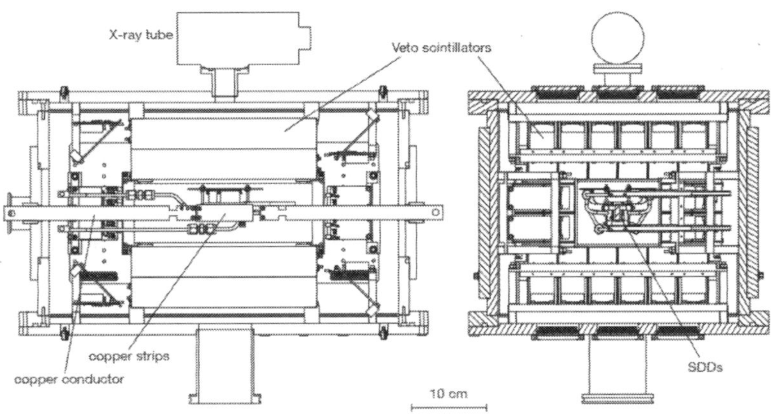

Figure 2. Schematic view of the VIP-2 apparatus installed inside the vacuum chamber.

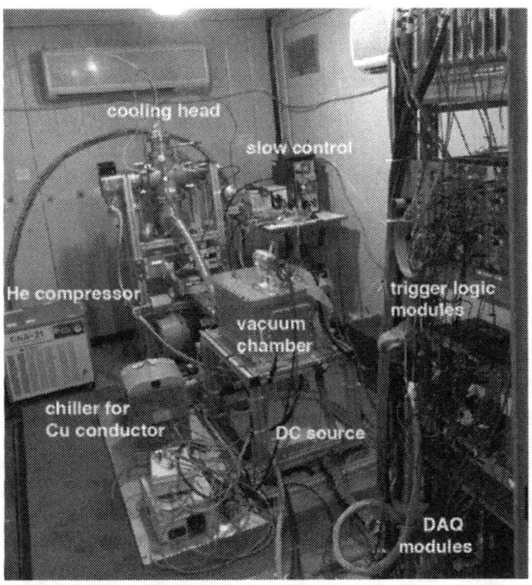

Figure 3. Picture of VIP-2 setup at the LNGS, before the shielding mounting. The apparatus in figure 2 is inside the vacuum chamber in the center of the image.

In November 2018, the setup was surrounded by a passive shielding complex composed of an external lead and inner copper layers, suppressing most of the background due to environmental radiation. More details on the VIP-2 setup, the trigger logic, data acquisition, and slow control can be found in [6, 10]. The VIP-2 Open Systems experimental setup is acquiring data in this final configuration (figure 3).

3 Analysis

The measured energy spectrum consists of a baseline and a Gaussian distribution for every emission peak (figure 4). The expected Kα emission signals from the Nickel (Ni) and the Copper (Cu) are 7478.2 and 8047.8 eV. The emissions have secondary lines, with a relative amplitude of 51% of the dominant one. These peaks are at 7460.9 and 8027.8 eV respectively for Ni and Cu, a relative shift of 17.3 and 20 eV from their main peak. For VIP-2 purposes, the "background" spectrum is the only signal expected without any current. A new Gaussian is expected when the current is injected: the X-ray emission from the PEP violating states. Its peak is expected at 7746.73 eV with the same width of Kα Cu since subjected approximately to the same electronic noise [6]. The signal's amplitude is proportional to the current injected since it increases the number of interactions electron-atom. However, it is expected to be negligible. Hence, the measurement aims to set the most stringent upper limit. Because of this delicate goal, we have different analysis approaches with both Frequentist and Bayesian inferences. In all methods, the selected energy region of interest for the analysis is from 7000 to 8500 eV, and we test deviation from the standard Kα transition in copper.

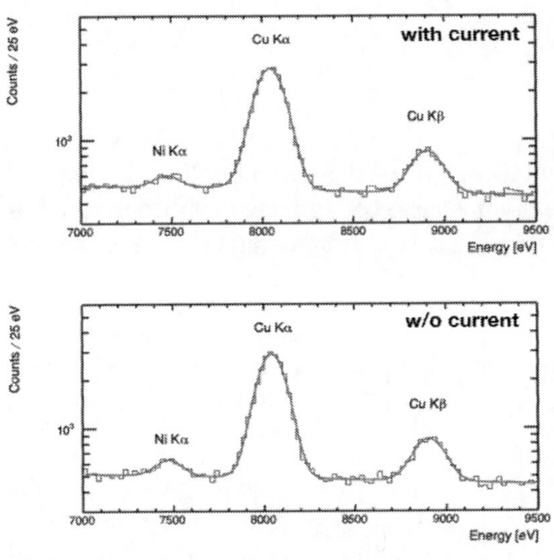

Figure 4. Spectrum of the signal in an energy region where the Kα from Cu and Ni and the Cu Kβ are visible. Above, with current; below without current and purely background. The red lines fit the spectra: the "with current" case also includes the Gaussian of the PEP-violation peaked at 7746.73 eV. See Sect. 3.1.2.

3.1 Frequentist Inference

Frequentist inference is the most classical approach. The paradigm is to extract the model parameters from counts, i.e., the best estimator of parameters θ given the "data" $E(\theta\,|\,\text{data})$. The "data" is assumed close enough to the asymptotic ideal distribution; hence, the inference holds for "large enough" statistics. We use two different methods: Spectra Subtraction (Sect. 3.1.1) and Simultaneous Fit (Sect. 3.1.2).

3.1.1 Spectra Subtraction

The most direct approach is simply subtracting the spectrum measured without (w/o) current from the one with current, both normalized by their time exposure (figure 5, left). If a PEP

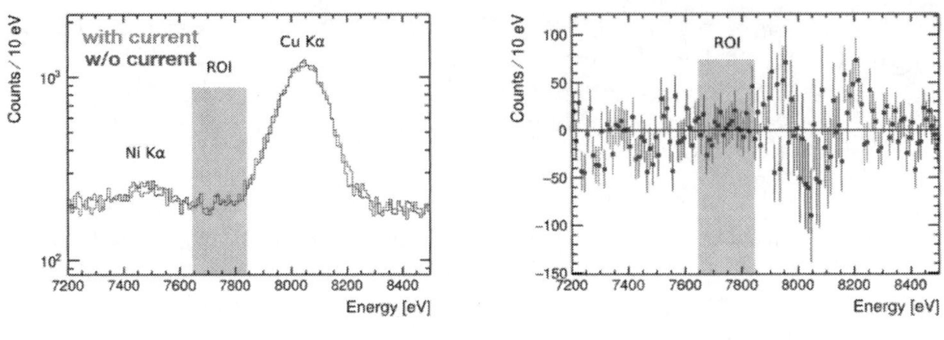

Figure 5. Left: normalized spectra with (red) and without (blue) current in comparison. Right: spectra difference "with" – "without" current. In both, the Region Of Interest is in evidence as a gray band.

violation occurs, some excess must be found in the Region Of Interest (ROI), where the signal is expected (figure 5, right). The ROI is defined as the Full-Width Half Maximum (FWHM) of the expected signal (i.e., the Cu Kα resolution) centered to the signal's peak: from 7647 to 7847 eV. From figure 5, the statistical uncertainty is too large to observe any excess in the ROI. However, this uncertainty lets us set an N_x and determine an upper limit to $\beta^2/2$. We typically show it at 99.7% Confidence Level (C.L.).

3.1.2 Simultaneous Fit

A more performing method consists of a simultaneous fit of the spectra collected with and without current. The two data are characterized by the same background shape, described as equation (2); the spectrum acquired with the current has one more component described by equation (3):

$$f_B(E) = p_0 + p_1 E + p_3 \left[e^{-\frac{(E-p_4)^2}{2p_5^2}} + 0.51\, e^{-\frac{(E-p_4-17.3)^2}{2p_5^2}} \right] + p_6 \left[e^{-\frac{(E-p_7)^2}{2p_8^2}} + 0.51\, e^{-\frac{(E-p_7-20)^2}{2p_8^2}} \right] \quad (2)$$

$$f_S(E) = p_9\, e^{-\frac{(E-7746.73)^2}{2p_8^2}} \quad . \quad (3)$$

All p_i are the free parameters of the fit, describing the baseline as a linear function and all the emissions as Gaussians (see Sect. 3): parameters describing positions of the standard transitions and resolutions are let be free to vary within the experimental uncertainties around the nominal values. To this end, a refined calibration procedure and the investigation of the relative systematic uncertainties are under finalization. Once the spectra are normalized by time exposure, simultaneous fit lets us measure p_9. Through it, we determine the $\beta^2/2$ upper limits, typically shown at 99.7% C.L..

3.2 Bayesian Inference

Unlike the frequentist one, this inference assumes ignorance on data abundance or scarcity. Hence, we look not for the best estimator of model parameters θ, but their probability distribution given "data": $P(\theta\,|\,\text{data})$, called the "posterior". From that, Bayes' theorem in our case is

$$P(S \mid N_{\text{data}}) = \frac{P(N_{\text{data}} \mid S) \cdot P_0(S)}{\int P(N_{\text{data}} \mid S) \cdot P_0(S)\, dS}. \tag{4}$$

$P(\text{data} \mid S)$ is the likelihood describing the model of the data distribution, the Poissonian

$$P(N_{\text{data}} \mid S) = \frac{(B+S)^{N_{\text{data}}} \cdot e^{-(B+S)}}{N_{\text{data}}!} \tag{5}$$

where B is the expected average of the background counts, S is the excess given by the PEP-violating signal and the searched parameter, and N_{data} are data counts acquired with the current where also the searched signal can be found. Finally, the $P_0(S)$ in equation (4) is called the "prior": an *a priori* hypothesis of weights for all possible parameter values. Since we know nothing about S except it cannot be lower than 0 or above previous experiments limit $S^{\max} = 574$ [6], a Heaviside $\theta(S - S^{\max})$ is considered. The final posterior, hence, become

$$P(S \mid N_{\text{data}}) = \frac{(B+S)^{N_{\text{data}}} \cdot e^{-(B+S)} \cdot \theta(S - S^{\max})}{\int_0^{S^{\max}} (B+S)^{N_{\text{data}}} \cdot e^{-(B+S)}\, dS}. \tag{6}$$

Equation (6) is a probability distribution, shown in figure 6 as an example; therefore, the estimator upper limit $\bar{S}$ at a given Confidence Level C.L. is the marginalized distribution

$$P(\bar{S}) = \int_0^{\bar{S}} P(S \mid N_{\text{data}})\, dS = \text{C.L.}. \tag{7}$$

S is the N_x in equation (1); hence, we use the resulting $\bar{S}$ to calculate the $\beta^2/2$ upper limit.

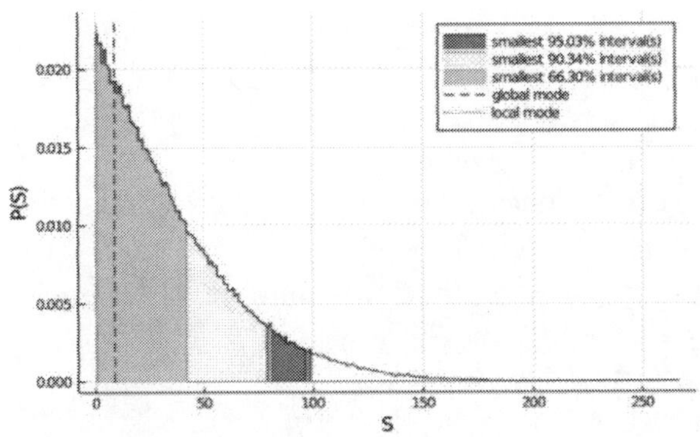

Figure 6. Example of a marginalized distribution for S. Green, yellow and red areas cover the 66.3, 90, and 95% of C.L.s. Data reported are for demonstration only and not meaningful.

The results of this analysis at 90% of C.L. are reported in [11] using data from the first light in the old VIP-2 configuration (see Sect. 2).

We are planning a different model for the present configuration and data set: it will be shown in future publications.

4 Discussion

We have discussed different analysis approaches: each has advantages and disadvantages. The *Spectra Subtraction* (Sect. 3.1.1) is the most direct and rough method, letting the analysis free from models or assumptions. However, it relies on the definition of an ROI and, more compromising, the statistics in this region: this is not a performing approach for the searched signal but more conservative. It can be used as the most conservative scenario of our results.

The *Simultaneous Fit* (Sect. 3.1.2) does not use any ROI but relies only on modeling the spectrum shape. We are improving its performance using advanced approaches like the CL_s method [12] (not discussed in this work), but the scarcity of the signal compared to the background spectra will always be a limiting factor.

While the Simultaneous Fit is the perfect methodology to cross-check the main approach, *Bayesian Inference* (Sect. 3.2) is our principal analysis method. It is excellent for low statistic signals because it relies more on its model, which is, in our case, simple and measurable. We are currently improving its performance through a different construction of the likelihood.

For the future, we are considering a Machine Learning method. There is a discussion on its potential in improving the results, architecture configuration, and what kind of input we can use (SDDs or background plus signal shapes).

Acknowledgement

This publication was made possible through the support of Grant 62099 from the John Templeton Foundation. The opinions expressed in this publication are those of the authors and do not necessarily reflect the views of the John Templeton Foundation. We acknowledge support from the Foundational Questions Institute and Fetzer Franklin Fund, a donor advised fund of Silicon Valley Community Foundation (Grants No. FQXi-RFP-CPW-2008 and FQXi-MGB-2011), and from the H2020 FET TEQ (Grant No. 766900) and INFN (VIP). We thank the Austrian Science Foundation (FWF) which supports the VIP2 project with the grants P25529-N20, project P 30635-N36 and W1252-N27 (doctoral college particles and interactions). K. P. acknowledges support from the Centro Ricerche Enrico Fermi - Museo Storico della Fisica e Centro Studi e Ricerche "Enrico Fermi" (Open Problems in Quantum Mechanics project).

References

[1] G. Lüders, B. Zumino, Phys. Rev. **110**, 1450 (1958)
[2] C. Curceanu, J. Gillaspy, R. Hilborn, Am. J. Phys. **80**, 561–577 (2012)
[3] A. Messiah, O. Greenberg, Phys. Rev. **136**, B248–B267 (1964)
[4] G. Gentile j, Il Nuovo Cimento **17**, 493–497 (1940)
[5] R. Amado, H. Primakoff, Phys. Rev. C **22**, 1338–1340 (1980)
[6] H. Shi, E. Milotti, S. Bartalucci, M. Bazzi, S. Bertolucci, A. Bragadireanu, M. Cargnelli, A. Clozza, L. De Paolis, S. Di Matteo et al. (VIP), Eur. Phys. J. C **78**, 319 (2018)
[7] E. Ramberg, G. Snow, Phys. Lett. B **238**, 438–441 (1990)
[8] M. Arzano, J. Kowalski-Glikman, Phys. Lett. B **760**, 69–73 (2016)
[9] A. P. Balachandran, T. Govindarajan, G. Mangano, A. Pinzul, B. Qureshi, S. Vaidya, Phys. Rev. D **75**, 045009 (2007)
[10] C. Curceanu (Petrascu), S. Bartalucci, S. Bertolucci, M. Bragadireanu, M. Cargnelli, S. Di Matteo, J. Egger, C. Guaraldo, M. Iliescu, T. Ishiwatari et al., Found. Phys. **41**, 282–287 (2011)
[11] K. Piscicchia, *et al.* (VIP), Entropy **22**, 1195 (2020)
[12] A. Read, Workshop on Confidence Limits **205** (2000)

Low-energy three-body collisions between an antiproton $\bar{\mathrm{p}}$ and muonic hydrogen atom H_μ

Renat A. Sultanov[1,*] and *S. K.* Adhikari[2,**]

[1]Odessa College, Department of Mathematics, 201 W. University Blvd., Odessa, TX 79764, USA
[2]Instituto de Física Teórica, UNESP – Universidade Estadual Paulista, 01140 São Paulo, SP, Brazil

Abstract. A few-body treatment is performed for two low-energy three-body collisions with participation of heavy charge particles such as an antiproton ($\bar{\mathrm{p}}$), deuterium (D), tritium (T) and a negative muon (μ^-). Specifically, the following reactions are considered: $\bar{\mathrm{p}} + (\mathrm{D}\mu^-)_{1s} \to (\bar{\mathrm{p}}\mathrm{D})_\alpha + \mu^-$ and $\bar{\mathrm{p}} + (\mathrm{T}\mu^-)_{1s} \to (\bar{\mathrm{p}}\mathrm{T})_\alpha + \mu^-$. The final state antinucleon-nucleon ($\bar{\mathrm{N}}\mathrm{N}$) interaction is included in these calculations and its influence on the cross sections and rates is estimated.

1 Introduction

Since $\bar{\mathrm{p}}$ is an anti-baryonic particle with the baryonic number $B = -1$, it would be interesting to discover the strong nuclear interaction between, for example, $\bar{\mathrm{p}}$ and a proton, i.e. the $\bar{\mathrm{p}}\mathrm{p}$ interaction in protonium (Pn) atom — a bound state of the particles: $\bar{\mathrm{p}}$ and p. This two-body system is also called as anti-protonic hydrogen. Additionally, it would be very useful and extremely interesting to consider, study and compare results between the following atomic systems: $\bar{\mathrm{p}}\mathrm{D}$ and $\bar{\mathrm{p}}\mathrm{T}$, where $\mathrm{D}{\equiv}^2\mathrm{H}^+$ is the deuterium nucleus and $\mathrm{T}{\equiv}^3\mathrm{H}^+$ is tritium. Perhaps it should be even worthier to obtain these atoms in their ground ($1s$) and close to ground states ($2s$, $2p$) when the systems are compact and nuclear interaction is effective.

It would be possible to prepare such atomic systems with the use of muons, for example, in the following three-body reaction:

$$\bar{\mathrm{p}} + (\mathrm{p}\mu^-) \to (\bar{\mathrm{p}}\mathrm{p})_\alpha + \mu^-, \tag{1}$$

where μ^- is a negative muon and α is the final atomic state of Pn. At low energy collisions $\alpha = 1s$, $2s$ or $2p$. This reaction has been considered in papers [1, 2]. It was found that the nuclear interaction in the final state is important [1]. Therefore, in the current paper the following three-body reactions are of our next special interest in the study of the antimatter-matter nuclear forces: $\bar{\mathrm{p}} + (\mathrm{D}\mu^-) \to (\bar{\mathrm{p}}\mathrm{D})_\alpha + \mu^-$, and $\bar{\mathrm{p}} + (\mathrm{T}\mu^-) \to (\bar{\mathrm{p}}\mathrm{T})_\alpha + \mu^-$.

In the next section we provide a brief description of our few-body approach based on the two-component Faddeev-Hahn-type equation formalism. We discuss the inclusion of the strong interaction in the framework of this equation. Sec. 3 represents our numerical results and Sec. 4 conclusions. Muonic atomic units (m.a.u.) are used in this work, i.e. $m_\mu = e^- = \hbar = 1$, where m_μ is the muon mass, e^- is the charge of an electron and $\hbar$ is Planck's constant.

*e-mail: rsultanov@odessa.edu; r.sultanov2@yahoo.com
**e-mail: sk.adhikari@unesp.br; adhikari44@yahoo.com

2 Few-body equations

In this work a detailed few-body approach is applied to the following three-charge-particle reactions at low energy collisions:

$$\bar{p} + (D\mu^-)_{1s} \rightarrow (\bar{p}D)_\alpha + \mu^-, \tag{2}$$

$$\bar{p} + (T\mu^-)_{1s} \rightarrow (\bar{p}T)_\alpha + \mu^-. \tag{3}$$

The method is based on the reduction of the total three-body wave function Ψ onto two Faddeev-type components [3]:

$$|\Psi\rangle = \Psi_1(\vec{\rho}_1, \vec{r}_{23}) + \Psi_2(\vec{\rho}_2, \vec{r}_{13}). \tag{4}$$

Eq. (4) reflects a fact that at low energy collisions, that is before the break-up threshold, only two asymptotic spatial configurations are possible in reactions (2) and (3). For example, in (2) the component $\Psi_1(\vec{\rho}_1, \vec{r}_{23})$ represents the $\bar{p}$+(Dμ) input channel, and the component $\Psi_2(\vec{\rho}_2, \vec{r}_{13})$ represents the output channel, i.e. $(\bar{p}D)_\alpha + \mu^-$. Vectors $(\vec{\rho}_k, \vec{r}_{ij})$ are so called Jacobi coordinates in the three-body system. These two configurations are shown in Fig. 1. Therefore, from works [4, 5] one can determine the Faddeev components by writing down the following set of two coupled equations:

$$\begin{cases} \left(E - \hat{T}_{\rho_1} - \hat{h}_{23}(\vec{r}_{23})\right)\Psi_1(\vec{r}_{23}, \vec{\rho}_1) = \left(V_{23}(\vec{r}_{23}) + V_{12}(\vec{r}_{12})\right)\Psi_2(\vec{r}_{13}, \vec{\rho}_2), \\ \left(E - \hat{T}_{\rho_2} - \hat{h}_{13}^{\bar{N}N}(\vec{r}_{13})\right)\Psi_2(\vec{r}_{13}, \vec{\rho}_2) = \left(\tilde{\mathbf{V}}_{13}(\vec{r}_{13}) + V_{12}(\vec{r}_{12})\right)\Psi_1(\vec{r}_{23}, \vec{\rho}_1). \end{cases} \tag{5}$$

Here: $V_{ij}(\vec{r}_{ij})$ are Coulomb potentials between the particles ($i \neq j$ = 1, 2, 3), $\tilde{\mathbf{V}}_{13}(\vec{r}_{13}) = V_{13}(\vec{r}_{13}) + v_{13}^{\bar{N}N}(\vec{r}_{13})$ and the two-particle target hamiltonians $\hat{h}_{23}(\vec{r}_{23}) = \hat{T}_{\vec{r}_{23}} + V_{23}(\vec{r}_{23})$ and $\hat{h}_{13}^{\bar{N}N}(\vec{r}_{13}) = \hat{T}_{\vec{r}_{13}} + V_{13}(\vec{r}_{13}) + v_{13}^{\bar{N}N}(\vec{r}_{13})$ with additional strong $\bar{p}$D (or $\bar{p}$T) nuclear potentials - $v_{13}^{\bar{N}N}(\vec{r}_{13})$. It is shown explicitly in these equations. $\hat{T}_{\vec{r}_{ij}}$ and $\hat{T}_{\vec{\rho}_k}$ are kinetic energy quantum-mechanical operators. The Faddeev decomposition avoids the over-completeness problems, because the subsystems are treated in an equivalent way in the framework of the two-coupled equations [6–8]. The correct asymptotes are guaranteed. The Faddeev-components are smoother functions of the coordinates than the total wave function $|\Psi\rangle$. The system of Eqs. (5) is equivalent to the Schrödinger equation – this is very important.

In order to solve Eqs. (5) a modified close-coupling approach is used. It leads to an expansion of the system's wave function components $|\Psi_1\rangle$ and $|\Psi_2\rangle$ into eigenfunctions $\varphi_n^{(1)}(\vec{r}_{23})$ and $\varphi_{n'}^{(2)\bar{N}N}(\vec{r}_{13})$ of the subsystem (target) Hamiltonians:

$$\begin{cases} \Psi_1(\vec{r}_{23}, \vec{\rho}_1) \approx \left(\displaystyle\int + \sum\right)_n f_n^{(1)}(\vec{\rho}_1)\varphi_n^{(1)}(\vec{r}_{23}), \\ \Psi_2(\vec{r}_{13}, \vec{\rho}_2) \approx \left(\displaystyle\int + \sum\right)_{n'} f_{n'}^{(2)}(\vec{\rho}_2)\varphi_{n'}^{(2)\bar{N}N}(\vec{r}_{13}). \end{cases} \tag{6}$$

This procedure brings a set of coupled one-dimensional integral-differential equations after the partial-wave projection [8]. The set of coupled integral-differential equations for $f_n^{(1)}(\vec{\rho}_1)$ and $f_{n'}^{(2)}(\vec{\rho}_2)$ can be solved in the framework of different close-coupling approximations, such as 2×1s, 2×(1s+2s), 2×(1s+2s+2p) etc. Symbol "2×" indicates that we use two independent sets of the target expansion functions.

The full potentials between $\bar{p}$ and D and $\bar{p}$ and T are more complex, because their second parts, $v_{13}^{\bar{N}N}(\vec{r}_{13})$, possess asymmetric $\bar{N}$-N nuclear interactions [9–12]. In this work we did not

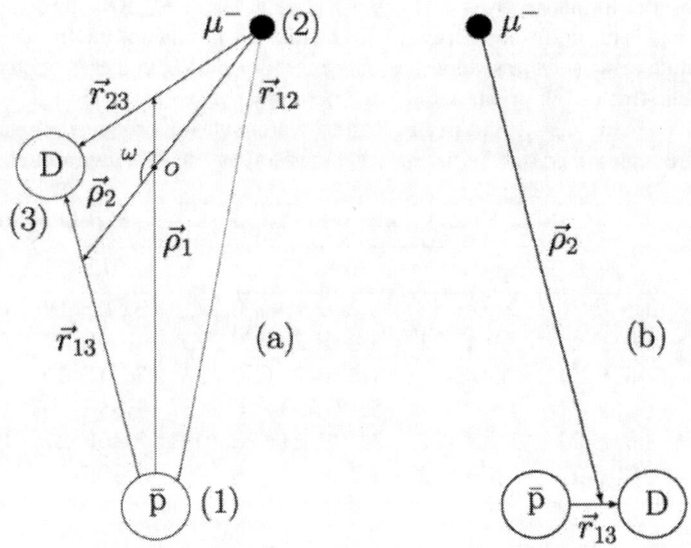

Figure 1. Two spatial coordinate configurations of the Coulomb three-body system p̄-μ-D. Here: (1) is an antiproton p̄, (2) is a negative muon μ^-, and (3) is the deuterium nucleus D. Vectors $\{\vec{r}_{23}, \vec{\rho_1}\}$ and $\{\vec{r}_{13}, \vec{\rho_2}\}$ are few-body Jacobi coordinates of the input and output channels of the reaction (2) correspondingly, O is the center of mass of the three-body system and ω is the angle between the Jacobi vectors $\vec{\rho_1}$ and $\vec{\rho_2}$.

explicitly include the strong interaction in our calculations, which is why in the case of the target p̄D and p̄T eigenfunctions we use pure two-body Coulomb (atomic) wave functions.

$$\varphi_{n'}^{(2)\bar{N}N}(\vec{r}_{13}) \approx \sum_{l'm'} R_{n'l'}^{(2)\bar{N}N}(r_{13})Y_{l'm'}(\hat{r}_{13}) \approx \sum_{l'm'} R_{n'l'}^{(2)}(r_{13})Y_{l'm'}(\hat{r}_{13}), \tag{7}$$

where $Y_{l'm'}(\hat{r}_{13})$ are spherical harmonics which depend on angular variables of $\vec{r}_{13}$, $\{n', l', m'\}$ are quantum atomic numbers and $R_{n'l'}^{(2)}(r_{13})$ are radial parts of the hydrogen-like atom wave function [13], for instance, the anti-protonic hydrogen atom. Nonetheless, the strong p̄D and p̄T interactions are approximately taken into account through the eigenstates $\mathcal{E}_{n'}$ which have shifted values from the original Coulomb levels $\varepsilon_{n'}$ [14], that is:

$$\mathcal{E}_{n'} \approx \varepsilon_{n'} + \Delta E_{n'}^{\bar{N}N} = -\mu_2/2n'^2 + \Delta E_{n'}^{\bar{N}N}, \tag{8}$$

where μ_2 is the reduced mass of the targets p̄D or p̄T in the output channels of the reactions (2) and (3).

In this work we apply Eqs. (5)-(8) with the use of an energy shift $\Delta E_{n'}^{\bar{N}N}$ in the eigenstate of p̄D and p̄T. The energy shifts can be computed, for example, with the use of the well-known Deser-Goldberger-Baumann-Thirring formula[14]:

$$\Delta E_{n'}^{\bar{N}N} = -\frac{4}{n'}\frac{a_s}{B_{\bar{p}H}}\varepsilon_{n'}, \tag{9}$$

where a_s is the pure nuclear strong interaction scattering length in the p̄+H collision (where H is D or T), i.e. without inclusion of the Coulomb interaction between the particles, $B_{\bar{p}H}$ is the Bohr radius of deuteronium, i.e. the p̄D atom or the p̄T one. Computational details of the Faddeev-Hahn equation formalism can be found in work [8].

Table 1. Total p̄D-atom formation cross sections $\sigma_{tr}(\varepsilon_{coll})$ and rates λ_D in the three-body reaction (2), where $\alpha = 1s$ and ε_{coll} is the collision energy (eV). The rate is a product of the formation cross sections and the corresponding center-of-mass velocities v_{cm} between p̄ and D_μ in the input channel. The results are represented in the framework of different close-coupling approximation models, Eq. (6). The cross section σ_{tr} is given in cm^2 and λ_D in m.a.u. Results with inclusion of the strong nuclear interaction between p̄ and D are represented only in the six-state model. Our rates λ_D are multiplied by factor of 5.

	$1s$	$1s$-$2s$	$1s$-$2s$-$2p$		(p̄-D Nucl.)	
ε_{coll}	σ_{tr}	σ_{tr}	σ_{tr}	λ_D	$\sigma_{tr}^{\bar{p}D}$	$\lambda_D^{\bar{p}D}$
0.001	2.82E-20	2.50E-20	1.22E-19	0.225	3.30E-19	0.612
0.01	8.93E-21	7.92E-21	3.84E-20	0.225	1.04E-19	0.611
0.1	2.83E-21	2.50E-21	1.20E-20	0.223	3.24E-20	0.601
0.5	1.26E-21	1.12E-21	5.18E-21	0.215	1.35E-20	0.560
1.0	8.91E-22	7.90E-22	3.50E-21	0.205	8.80E-21	0.515
2.0	6.30E-22	5.57E-22	2.26E-21	0.188	5.30E-21	0.439

In the literature one can find other theoretical formulas to compute $\Delta E_{n'}^{\bar{N}N}$, for instance, in papers [15, 16]. In future calculations it would be interesting to apply some of these theories together, for example, with the inclusion of relativistic effects in the heavy p̄D and p̄T atoms [17, 18].

Since muon is ~207 times heavier than e^-, the muonic hydrogen atom H_μ has a very small size. Therefore, in reactions (2)-(3) antiproton can very closely approach H_μ. However, annihilation between p̄ and D/T will be significantly prevented because of the μ^- screening effect and a strong p̄ and μ^- Coulomb repulsion. The quantum-mechanical p̄ tunneling through the muonic-atomic orbit H_μ is also significantly suppressed. This fact can be seen from the following quantum-mechanical tunneling probability formula [13]:

$$B = \exp\left\{-\frac{2}{\hbar}\int_0^{\rho_0}\sqrt{2M(U(r)-E)}dr\right\}, \tag{10}$$

where B is the probability, E is the total energy in the three-body system, M is the p̄-H_μ reduced mass, and $U(r)$ is the interaction potential between p̄ and H_μ:

$$U(r) = \left(\frac{1}{r} + \mu_0\right)e^{-2\mu_0 r}, \tag{11}$$

μ_0 is the muonic hydrogen reduced mass, i.e. $\mu_0 \approx 207m_e$, where m_e is the electron mass. The integration in the Eq. (10) can be done up to $\rho_0 \approx 10$ m.a.u. One can compute the integral (10) and show that the argument of the exponent in Eq. (10) is a large number. Therefore, in the first-order approximation p̄ tunneling can be neglected. In the case of similar atomic system, where one has an electron e^- instead of muon, p̄ can easily penetrate through the light e^- cloud and annihilate with the hydrogen isotopes before the three-body reaction occurs.

3 Results

This section represents our results for reactions (2) and (3). As in paper [1], in the current work we carried out numerical calculation for the transitions to the ground states of the antiprotonic atoms, i.e. $\alpha = 1s$ in Eqs. (2)-(3). In this state the size of the p̄D and p̄T atoms

Table 2. Total p̄T-atom formation cross sections $\sigma_{tr}(\varepsilon_{coll})$ and rates λ_T in the three-body reaction (2), where $\alpha = 1s$ and ε_{coll} is the collision energy (eV). The rate is a product of the formation cross sections and the corresponding center-of-mass velocities v_{cm} between p̄ and T_μ in the input channel. The results are represented in the framework of different close-coupling approximation models, Eq. (6). The cross section σ_{tr} is given in cm^2 and λ_T in m.a.u. Results with inclusion of the strong nuclear interaction between p̄ and T are represented only in the six-state model. Our rates are multiplied by factor of 25.

	$1s$	$1s$-$2s$	$1s$-$2s$-$2p$		(p̄-T Nucl.)
ε_{coll}	σ_{tr}	σ_{tr}	σ_{tr}	λ_T	$\lambda_T^{\bar{p}T}$
0.001	2.23E-20	2.64E-20	7.80E-20	0.685	1.315
0.01	7.04E-21	8.34E-21	2.45E-20	0.685	1.316
0.1	2.23E-21	2.64E-21	7.82E-21	0.687	1.322
0.5	9.95E-22	1.18E-21	3.55E-21	0.698	1.352
1.0	7.03E-22	8.36E-22	2.56E-21	0.711	1.391
2.0	4.96E-22	5.93E-22	1.89E-21	0.742	1.473

is about ~10 fm. Therefore, one can expect that the contribution of the nuclear forces to the three-body scattering cross sections and rates will be significant. These reactions, in our opinion, can be useful to study nuclear matter-antimatter interactions. Table 1 shows our results for reaction (2). Specifically, the table depicts our p̄-transfer cross-sections $\sigma_{tr}(E)$ and corresponding reaction rates λ_D at low energy collisions. The results are shown in the framework of the $2\times1s$, $2\times(1s+2s)$ and $2\times(1s+2s+2p)$ close-coupling approximation. One can see that the contribution of the $2p$ atomic states is large–slow p̄ can approach the Dμ atom particularly close and significantly polarize it. Therefore, the inclusion of the polarization channels in Eq.(6) is important. Finally, taking into account the nuclear interaction between p̄ and D increased the cross-section almost three times. This result can be seen from Table 1: from the cross section column, $\sigma_{tr}^{\bar{p}D}$, and from the corresponding reaction rate one, i.e. $\lambda_D^{\bar{p}D}$.

The inclusion of this interaction has been done within the framework of Eqs. (7)-(8) and (9). From Ref. [19] (Table 5.2) one can find the results of the p̄D atom ground-state Coulomb energy shifts. The averaged experimental result has the following value:

$$\Delta E_{n'=1s}^{\bar{p}D} = 1050 \text{ eV}. \tag{12}$$

This result was adopted in this work. However, the result for $\Delta E_{n'=1s}^{\bar{p}D}$ differs quite significantly from the theoretical calculations based on the three-body Faddeev theory [12] and another older work [11]. With the use of the given $\Delta E_{n'=1s}^{\bar{p}D}$ (12), one can estimate the p̄+D scattering length:

$$a_s^{\bar{p}D} = -\frac{B_{\bar{p}\mathrm{H}}}{4\varepsilon_{n'=1s}}\Delta E_{n'=1s}^{\bar{p}D} = 0.682 \text{ fm}, \tag{13}$$

and compute the Coulomb energy shift for 2s and 2p states ($n' = 2$):

$$\Delta E_{n'=2}^{\bar{p}D} = 131.25 \text{ eV}. \tag{14}$$

All these values have been included in the Eqs. (5) in our calculation of the reaction (2).

Table 2 shows our results for the collision (3). This is a very attractive and special three-charge-particle reaction with participation of tritium (T), which is a radioactive hydrogen isotope. It would be extremely interesting to investigate the influence of the strong interaction between p̄ and T on the rate of the reaction (3). With the use of this reaction it should be possible to study nuclear interaction between p̄ and radioactive T.

In order to carry out calculations of the process (3) one needs the nuclear energy shifts of the ($\bar{p}$T) Coulomb levels as an input. However, to our best knowledge, this data are not available in the literature so far. Therefore, at this time it is not possible to carry out quality and strict calculations for this reaction. Nonetheless, we will depict our preliminary results for the reaction rates of (3) with the use of a model approach. Table 2 shows our reaction cross-sections $\sigma_{tr}(E)$ and rates λ_T. One can see that the inclusion of the $2p$ atomic states is important for the transfer channel. Also, the transfer cross section is smaller than the corresponding transfer cross-section in the reaction (2). This result agrees with the results of previous paper [2].

In regard to the nuclear interaction in (3) we also use Eqs. (7)-(9) in order to estimate the effect of the strong $\bar{p}$T interaction. However, as we mentioned above, there are no results for the $\bar{p}$T ground-state Coulomb level energy shifts, i. e. no results for $\Delta E^{\bar{p}T}_{n'=1s}$. We have come across only a paper where we found information about $\bar{p}$+T scattering [20]. Therefore, in the current paper, we apply a model approach for the Coulomb level energy shift $\Delta E^{\bar{p}T}_{n'}$. Note that in the case of the two-particle system ($\bar{p}$p) the energy shift is $\Delta E^{\bar{p}p}_{n'=1s} = 540$ eV. In the case of the three-nuclear-particle system $\bar{p}$D: $\Delta E^{\bar{p}D}_{n'=1s} = 1050$ eV. Therefore, we assume that in the case of the four-particle system, i.e. $\bar{p}$T, the ground state Coulomb level energy shift due to the nuclear interaction is $\Delta E^{\bar{p}T}_{n'=1s} \approx 1575$ eV.

We adopt this value and use it in order to complete our calculation of the rate of the reaction (3) with the inclusion of the strong $\bar{p}$T interaction in the final state. As a preliminary treatment, the energy range for this calculation was only: 10^{-3} eV $\leqslant E \leqslant 2$ eV. We can see from Table 2 that the nuclear $\bar{p}$T interaction effect is large. In order to compare our results with the results of paper [2] our $\bar{p}$+(Tμ^-) rates λ_T and $\lambda_T^{\bar{p}T}$ were multiplied by 25.

4 Conclusions

The main goal of this paper is to carry out calculations of the charge-transfer reactions (2) and (3) at low energies, and investigate the influence of the strong nuclear interaction on the final states of these reactions, i.e. the influence of the $\bar{p}$D and $\bar{p}$T short-range nuclear forces on the output scattering parameters. A set of coupled two-component Faddeev-Hahn-type equations has been applied together with a modified close-coupling approximation technique. The main advantage of this three-body method over other few-body adiabatic approaches is that we utilize an independent formulation of the two-body targets in the reactions (2) and (3): for example, the (Dμ) atom in the input channel and the ($\bar{p}$D) atom in the output channel. This distinctive property of the Eqs. (5)-(6) allows us to avoid the over-completeness problems and provide accurate three-body asymptotes for $\Psi_1(\vec{r}_{23},\vec{\rho}_1)$ and $\Psi_2(\vec{r}_{13},\vec{\rho}_2)$. The set of coupled Eqs. (5) is equivalent to the Schrödinger equation. We treat the Coulomb three-body systems as systems with arbitrary masses, i.e. the masses of the charged particles are taken as they are. We do not apply any type of adiabatic approximation, when the dynamics of heavy and light parts of a system are separated.

In the case of the reaction (2) it is shown that the effect of the final-state nuclear interaction is significant - up to 275%. Therefore, one could conclude that the three-body reaction could be considered as a possible candidate for future experiments with participation of low-energy antiprotons $\bar{p}'s$ and muons in order to produce the antiprotonic hydrogen atom ($\bar{p}$D) and study the nuclear interaction between $\bar{p}$ and D at low energies. In regard to the very interesting ($\bar{p}$T) system we carried only preliminary estimations of the influence of the strong $\bar{p}$T forces on the reaction outputs. It was also found that the contribution is large.

In conclusion, we would like to make a cautious assumption, that the low-energy collisions between antiprotons and muonic atoms could also be useful to study a quite old problem in nuclear physics concerning the long-range part of the strong $\bar{N}$N interaction [21].

Acknowledgements

We wish to acknowledge the support from CNPq and FAPESP of Brazil.

References

[1] R.A. Sultanov, D. Guster, S.K. Adhikari, Atoms **6**, 18 (2018).
[2] A. Igarashi, N. Toshima, Eur. Phys. J. D **46** (2008) 425.
[3] L.D. Faddeev, Zh. Eksp. Teor. Fiz. **39**, 1459 (1960) [Sov. Phys. JETP **12**, 1014 (1961)].
[4] Y. Hahn, K. Watson, Phys. Rev. A **5**, 1718 (1972).
[5] Y. Hahn, Nucl. Phys. A**389**, 1 (1982).
[6] R.A. Sultanov, S.K. Adhikari, Phys. Rev. A **61**, 227111 (2000); **62**, 022509 (2000).
[7] R.A. Sultanov, D. Guster, J. Comp. Phys. **192**, 231 (2003).
[8] R.A. Sultanov, D. Guster, J. Phys. B: At. Mol. Opt. Phys. **46**, 215204 (2013).
[9] C.B. Dover, J.-M. Richard, Phys. Rev. C **25**, 1952 (1982).
[10] J.-M. Richard, M.E. Sainio, Phys. Lett. B **110**, 349 (1982).
[11] S. Wycech, A.M. Green, J.A. Niskanen, Phys. Lett. B **152**, 308 (1985).
[12] G.P. Latta, P.C. Tandy, Phys. Rev. C **42**, 1207 (1990).
[13] L.D. Landau, E.M. Lifshitz, *Quantum-Mechanics (Non-relativistic Theory), Course of Theoretical Physics, Vol. 3* (3rd Ed., Butterworth-Heinemann, 2003).
[14] S. Deser, M.L. Goldberger, K. Baumann, W. Thirring, Phys. Rev. **96**, 774 (1954).
[15] T. L. Trueman, Nucl. Phys. **26**, 57 (1961).
[16] V.S. Popov, A.E. Kudryavtsev, V.D. Mur, Sov. Phys. JETP **50**, 865 (1979).
[17] T. Ueda, Prog. Theor. Phys. **62**, 1670 (1979).
[18] J. Thaler, J. Phys. G: Nucl. Phys. **9**, 1009 (1983); **11**, 201 (1985).
[19] E. Klempt, F. Bradamante, A. Martin, J.-M. Richard, Phys. Rep. **368**, 119 (2002).
[20] L.A. Kondratyuk, M. Z. Shmatikov, Sov. J. Nucl. Phys. **38**, 216 (1983).
[21] W.W. Buck, C.B. Dover, J.-M. Richard, Ann. Phys. **121**, 47 (1979).

AUTHOR INDEX

AUTHOR INDEX

AUTHOR INDEX

EDP Sciences – Web of Conferences
17, Avenue du Hoggar
Parc d'Activité de Courtabouf
BP 112
F-91944 Les Ulis Cedex A
France

ISBN 978-1-7138-5228-5